CANTILEVER ARCHITECTUR

Architects are often fascinated by dramatic cantilevers. Indeed, cantilevers are widely used in architecture for various reasons and in different scales. This book is organized to present studies on the entire range of cantilevers employed in architecture: cantilevered furniture, cantilevered building components and significantly cantilevered major building parts, all primarily subjected to gravity loads; and finally a building as a whole, including supertall buildings of complex-shapes, as a vertical cantilever against lateral loads. Each chapter presents a specific subject on either horizontal or vertical cantilevers of different scales, functions and forms based on the following outline: brief historical review, basic structural principles, and holistic analysis of real world or theoretical examples. By presenting almost all different cantilever design scenarios in architecture, this book is a unique and essential reference on cantilever architecture.

Kyoung Sun Moon, PhD, AIA, is Associate Professor at Yale University School of Architecture. Educated as both an architect and engineer, his primary research area is integration between the art and science/technology of architecture, with a focus on tall and other structurally challenging buildings. Prior to joining the Yale faculty, he taught at the University of Illinois at Urbana-Champaign and worked at Skidmore, Owings, and Merrill in Chicago and the Republic of Korea Navy. He received his PhD from Massachusetts Institute of Technology.

CANTILEVER ARCHITECTURE

KYOUNG SUN MOON

Routledge
Taylor & Francis Group

NEW YORK AND LONDON

First published 2019
by Routledge
711 Third Avenue, New York, NY 10017

and by Routledge
2 Park Square, Milton Park, Abingdon, Oxon, OX14 4RN

Routledge is an imprint of the Taylor & Francis Group, an informa business

Library of Congress Cataloging-in-Publication Data
A catalog record for this title has been requested

ISBN: 978-1-138-67418-9 (hbk)
ISBN: 978-1-138-67421-9 (pbk)
ISBN: 978-1-315-56144-8 (ebk)

Typeset in Univers
by Florence Production Ltd, Stoodleigh, Devon, UK

CONTENTS

FOREWORD

CANTILEVERS ABOUND in nature in many morphological and tectonic forms. Trees and mountains are the best examples of vertical cantilevers in nature that resist lateral wind forces. Historically, buildings and bridges have been using cantilevers to create projections that are employed either for structural or aesthetic reasons, or for both. With rapid advances in material and construction technologies during the past several decades, cantilevers are now appearing in buildings all around the world, in limitless forms and combinations. The design and engineering of cantilevers is also currently one of the utmost exhilarating, cutting edge, and vigorous fields of research in industry in various contexts.

Horizontal cantilevers in buildings are a very common sight. Interestingly, tall buildings are truly colossal vertical cantilevers by themselves spiking out of the ground towards the sky and are subjected to the fierce lateral forces of high winds and earthquakes. Even in building foundations, engineers use cantilevers to their advantage for different types of footings. Literature on cantilevers flourishes in scattered forms; architects and engineers employ them in solving many simple and complex problems in design offices and studios. Yet surprisingly, there is no book that explicitly and entirely focuses on the cantilever, which is becoming so popular in the modern world. Indeed, a book on this subject has been long overdue. *Cantilever Architecture* by Kyoung Sun Moon bridges that gap.

With architectural research currently dominated by the continuing structural emphasis on verticality and the race for the tallest construction, it seems timely now to focus on the unique and critical importance and value of the under-researched but increasingly germane architectural properties, possibilities, and challenges pertaining to cantilevers. The interesting features about the cantilevers are that they are aesthetically pleasing – often with sculptural qualities – endowed with quasi-dynamic forms that seem to fly in the air in static motion; they can minimize obstructive support structures and offer desirable versatility of application. They can be strategically employed in long-span structures to reduce bending and deflection of main spans. And their application is not limited to only buildings and bridges, but also to many industrial, mining, telecommunication and aviation structures.

Why is this book important? In partial answer to this question, others in the future will likely write books on this topic – but few will have the qualifications of Kyoung Sun Moon. He is uniquely qualified to write this book as he is an architect, a structural engineer and a dedicated researcher. He has thoroughly dug into this subject and produced this generously illustrated book offering some fresh perspectives. No wonder his motivation to write this book stems from his keen observation of different types and applications of cantilevers. Because of his combined architectural and engineering background he has been able to successfully examine the technical aspects of cantilever systems and express them in qualitative terms.

In *Cantilever Architecture* Kyoung Sun Moon has presented an informative and refreshing account of fascinating exploration of cantilevers in lucid and understandable language. The work will be of great value to those in the architectural and engineering professions, as well as to students aspiring to learn in depth the intricate details of cantilevers. It demands careful study.

Mir M. Ali
Professor Emeritus of Architecture
University of Illinois at Urbana-Champaign

INTRODUCTION

ARCHITECTS ARE OFTEN FASCINATED by the use of support-free structures to solve certain design problems. They often want a significant portion of their buildings to be supported without conventional columns or walls so the buildings have a feel of hovering. Two typical design approaches for the situation are direct horizontal cantilevering and hanging from the top. The latter, using tensile members, also usually requires a substantial cantilever structure at the top to hang the desired portion of the building from there.

Cantilevers have been used in buildings throughout the history of architecture. However, the scale of cantilevers was limited by the properties of traditional building materials. Supported by the development of stronger and stiffer modern structural materials, such as steel and reinforced concrete in the 19th century, and continued advancements of construction techniques, dramatic cantilevers of unprecedented scales began to emerge. Today, many buildings throughout the world find their design solutions using large cantilevers.

Despite many architects' enthusiasm and fascination, publications on systematic studies of cantilever architecture are very limited. Designing buildings with large cantilevers requires significant structural engineering considerations, which are typically beyond architects' capability. This book is to help practicing architects and architecture students, especially at the early stages of design, conceptually understand how cantilever architecture of many different configurations works structurally and how it can be better integrated synergistically with architectural and other design aspects.

Cantilevers are used in architecture for various reasons and in different scales. A significant portion of primary building structures can be cantilevered to produce more dramatic sculptural expressions, or, very practically, to maximize occupiable space using air rights beyond the property limit, or for many other design-specific reasons. Smaller scale cantilevers are also used as building components, such as cantilevered entrance canopies, balconies, stairs, etc. In addition, furniture in buildings is often designed with cantilevers, such as cantilevered chairs and tables. Chapter 1 through Chapter 3, which compose Part I of this book, present various systems and their load carrying mechanisms of cantilever structures of many different functions, scales and configurations, in relation to their architectural and other design-related issues.

Furniture performs as an integral part of interior space. Many architects have been attracted by the idea of designing furniture customized for their own designed buildings or for mass production. Chapter 1 presents how cantilevered furniture performs, and what the performance differences are between cantilevered and non-cantilevered furniture. Performances of cantilevered chairs and tables of various configurations are comparatively studied with real world examples designed by many architects and designers such as Mart Stam, Mies van der Rohe, Marcel Breuer, Hans Luckhardt, Garrit Rietveld, Eileen Grey, to name but a few.

As constructed objects, buildings are composed of various physical components of different scales. Chapter 2 presents how cantilevered major building components, such as stairs, balconies and canopies, are designed, constructed and perform. Cantilever stairs of various examples are studied from ancient stone cantilever stairs to those of modern building materials such as reinforced concrete, steel and glass. Cantilevered balconies and canopies of different configurations and materials are studied. Not only structural but also architectural, environmental and other design related aspects are discussed holistically with real world examples.

Chapter 3 presents the concept of cantilever employed for a significant portion of a building. It begins with discussions on efficient proportioning of symmetrical and asymmetrical cantilevers. After that, cantilevered buildings are categorized for systematic studies based on their configurational characteristics, such as large one-sided cantilevers, two-sided cantilevers, merged cantilevers and stacked multiple cantilevers. For each category, its basic structural concept and performance are introduced first and its applications to built examples are studied. In many cases, alternative design scenarios are comparatively studied to simulate typical real world design processes and, consequently, help architects better understand how buildings with large cantilevers of alternative configurations perform differently.

The term, cantilever, in architecture, is typically used for horizontal structures supported at only one end to carry primarily gravity loads. However, buildings are subjected to not only gravity but also lateral loads, such as wind and seismic loads. Regarding lateral loads, any building is considered as a vertical cantilever supported on the ground, and the issue of a building as a vertical cantilever becomes more important as a building becomes taller. While Part I of this book presents systematic studies on horizontal cantilevers primarily subjected to gravity loads, Part II composed of Chapter 4 through Chapter 6 is devoted to vertical cantilevers against lateral loads.

In order to produce more efficient vertical cantilevers, it is important to maximize structural depths of buildings against lateral loads. Chapter 4 presents various lateral load resisting systems for vertical cantilevers divided into three different conceptual categories – interior structures, exterior structures and interior-exterior-integrated structures. Performance characteristics of different structural systems in each category are investigated in relation to architectural and other design-related issues, theoretically and with real world examples. Furthermore, comparative performances between the systems within each category as well as between the categories are discussed.

As buildings become ever taller and more slender, lateral vibrations due to wind loads may cause serious occupant discomfort and serviceability issues. Structures with more damping dissipate the vibration energy more quickly, and, consequently, reduce structural motions more rapidly. Chapter 5 presents various damping strategies for vertical cantilevers. A rigorous and more reliable increase in damping could be achieved by installing auxiliary damping devices integrated with the primary structural system. Both passive and active systems are studied. An emphasis is placed on the passive system which is further categorized into material and mass dampers.

Early design of tall buildings culminated with the dominance of the International Style, which prevailed for decades and produced prismatic Miesian style towers all over the world. Today's pluralism in architectural design has produced tall buildings of many different forms, including more irregular and complex forms. Chapter 6 presents dynamic interactions between the various complex building forms and structural design of tall buildings. Complex building forms are categorized into twisted, tilted, tapered and free forms, and structural performances of these complex-shaped vertical cantilevers are studied in conjunction with architectural and other design-related issues. In addition, various design approaches, performances and future potentials of conjoined towers – another recently emerging tall building type – are discussed.

This book is organized to present studies on the whole range of cantilevers employed in architecture. Each chapter presents a specific subject based on the following outline: brief historical review; basic principles; holistic analysis of real world examples. By presenting almost all different cantilever design scenarios in architecture, this book is expected to be a unique and essential reference on cantilever architecture.

PART 1
HORIZONTAL CANTILEVERS

CHAPTER 1
CANTILEVERED FURNITURE

FURNITURE PERFORMS AS A VERY IMPORTANT integral part of interior space. Many architects have been fascinated by the idea of designing customized furniture for their own designed buildings. Frank Lloyd Wright, one of the greatest architects of all time, designed furniture as unified components of his architecture. He customized this furniture for many of his designed buildings to integrate the design of the entire building as a whole. For many early modernism architects, it was almost necessary to design furniture for their own buildings because of the stylistic lag in traditional furniture design compared to the buildings they designed with new materials, technology and design ideas. Though the early modernists, such as Ludwig Mies van der Rohe, Le Corbusier, Alva Alto and Marcel Breuer, to name a few, initially designed furniture mostly for their own designed buildings, their furniture was typically intended to be mass-produced by machines as well.

Cantilevered chairs, as a pioneer of employing the principle of significantly proportioned cantilever in furniture, were brought to the world in this context. The concept of cantilever has a long history in architecture with smaller scales and proportions in the past with the use of traditional building materials such as stone and wood. Cantilevers of significant proportions only became possible with stronger and stiffer modern structural materials such as steel and reinforced concrete. In furniture design, the inspiring initial use of tubular steel by Marcel Breuer and the original cantilevered frame designed for a chair by Mart Stam using gas pipes in the 1920s were put together to produce the early versions of cantilever chairs and numerous variations soon after. Since then, the cantilever chair using tubular steel and later on other materials, has become a very important category of contemporary chairs.

This chapter presents how cantilevered furniture, such as cantilevered chairs and tables work structurally and what the performance differences are between cantilevered and non-cantilevered furniture. Further, cantilevered furniture items of various configurations are comparatively studied with real world examples designed by architects and designers, such as Mart Stam, Mies van der Rohe, Marcel Breuer, Garrit Rietveld, Eileen Grey, to name but a few.

1.1 CANTILEVER CHAIRS

Chairs, which intimately hold people who sit on them, are one of the most popular designed furniture items by architects. The essential structural configuration of most common chairs includes legs which support seats and vertically cantilevered backrests. Figure 1-1 shows a typical simple chair composed of the minimum necessary elements. The seat of the chair is supported by the four vertical corner legs. In order to provide lateral stability, the legs and the horizontal framing members of the seat should be rigidly connected. The seat can perform as a diaphragm which stabilizes the horizontal frames of the seat. The vertical backrest should be designed and constructed as a vertical cantilever so it can resist horizontal force exerted by leaning actions of the user. Numerous variations from this minimal configuration are possible and available.

When Mart Stam conceived a cantilever chair in 1924, it was a radically different and unprecedented chair configuration. The structure of the cantilever chair is composed of continuous pipes as can be seen in the prototype shown in Figure 1-2. In terms of the chair structure's basic composition, the two rear legs of the simple chair shown in Figure 1-1 are removed, while the front legs

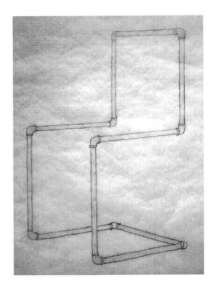

Figure 1-1. Typical simple chair. Design by Shenxing Liu for Greenington Fine Bamboo Furniture.

Figure 1-2. Sketch of the original prototype of the cantilever chair structure by Mart Stam.

are horizontally extended to the base location of the removed rear legs. Simply removing the rear legs without this extension would make the chair unstable. The two side frames which define the profile of the cantilever chair are connected at the top and bottom to complete the three-dimensional loop type structure for stability. With regard to the materials for the chair structure, metal pipes were used. The loop structure was produced by connecting 10 straight gas pipe members with 10 plumbing elbows. Once the seat and backrest of desired materials, such as leather or wicker, are placed between the two metal pipe side frames, the cantilever chair is completed.

The production of actual cantilever chairs never followed the original construction strategy with multiple pieces of metal pipes and elbows. Instead, seamless polished tubular steel was employed for the frames of the cantilever chairs. Nonetheless, since the introduction of the cantilever chair concept by Mart Stam, this has opened an important new category of modern furniture design. Many architects and designers attracted by the cantilever concept have designed many different cantilevered chairs of various configurations and other cantilevered furniture pieces.

Figure 1-3. Cantilever Chair by Mart Stam (left) and MR10 by Ludwig Mies van der Rohe (right). With permission of Casa Factory (L), 1stdibs (R).

Figure 1-3 shows the cantilever chair by Mart Stam and that by Ludwig Mies van der Rohe. The overall configurations of Stam's and Mies' cantilever chairs are similar, except that the front legs of Mies' cantilever chair have a semicircular form. The frames of the both chairs are made of continuous polished tubular steel and consequently all connections can be considered as rigid connections. In these designs, gravity loads applied to the seats are carried by cantilever actions of the frames.

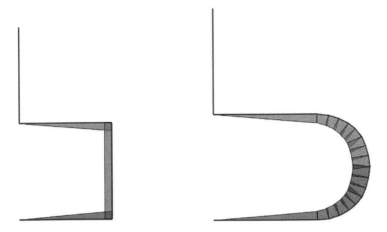

Figure 1-4. Bending moment diagrams of cantilever chairs by Mart Stam (left) and Ludwig Mies van der Rohe (right) subjected to gravity loads on the seats.

Figure 1-4 shows the bending moment diagrams of the two cantilever chairs by Stam and Mies with only gravity loads applied to the seats. Mies' cantilever chair, which has longer cantilever due to the semicircular front legs, develops larger bending moments when the identical load is applied to the seats. Figure 1-5 shows exaggerated deformed shapes of the two cantilever chairs. Mies' chair with longer cantilever and greater bending moments produces a larger deformation. As long as the chair is strong enough to carry the applied loads, this larger deformation can perhaps be better for the purpose of the chair in terms of providing comfortable sitting experience because, if it is not excessive, a larger deformation can work as an added springy cushion.

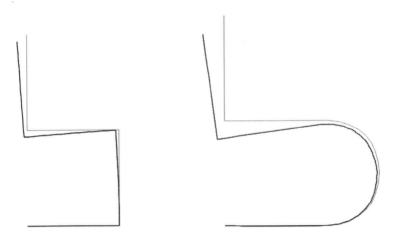

Figure 1-5. Deformed shapes of cantilever chairs by Mart Stam (left) and Ludwig Mies van der Rohe (right) subjected to gravity loads on the seats.

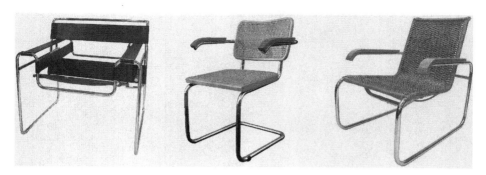

Figure 1-6. Marcel Breuer's B3 (Wassily), B64 (Cesca) and B35.
With permission of 1stdibs.

However, in building structures, which will be discussed extensively in the remaining chapters of this book, larger deformations due to greater cantilever lengths and bending moments are typically a serious structural design issue for serviceability.

Marcel Breuer is one of the most influential designers in developing furniture framed with polished tubular steel. He used this material first for home furniture in his chair B3, Wassily, in 1925, right before the production of the cantilever chair by Mart Stam. Marcel Breuer also designed his version of cantilever chairs, such as B32 and B64, Cesca, in 1928. While Mart Stam and Mies van der Rohe used flexible materials for the seats and backrests of their polished tubular steel cantilever chairs, Marcel Breuer combined wood and wicker as can be seen in Figure 1-6.

The configuration of Breuer's tubular steel frame cantilever chair is basically the same as that of Stam's except for the different material choice for the seats and backrests. In his cantilevered armchair, B64, the continuous seat and backrest frame members are cantilevered from the base frame members which rest on the floor, and the armrests are secondarily cantilevered from the backrest frames. However, in his B35, Breuer further developed the concept of the cantilever chair to a more comfortable and luxurious version. In B35, the continuous seat and backrest frame members and the armrest frame members are, in a sense, independently cantilevered to the opposite directions from the more horizontally elongated base frame members. The elongated base frame members allow a deeper seat and taller backrest. Furthermore, the seat and backrest are more reclined in B35 than in B64, for superior comfort.

Marcel Breuer designed not only cantilever chairs but also a cantilever sofa, F40, for multiple users as can be seen in Figure 1-7. Compared to B64 or B32 (Breuer's cantilever chair with no armrest), the height-to-width aspect ratio of the main frame is much smaller in F40. Reduced height of the front leg frame members and increased length of the base and seat frame

11

Figure 1-7. Marcel Breuer's cantilevered sofa, F40. With permission of 1stdibs.

members provide much greater safety against overturning failure which could be possible by leaning actions of the users towards the back of the sofa. The reduced height means a shortened overturning moment arm, while the increased length means an increased resisting moment arm. Padded cushions are used for the seats and backrests in F40 to provide enhanced comfort.

Though not directly related to cantilever chair designs, Marcel Breuer also powerfully used the concept of cantilever in his buildings. In the Whitney Museum of American Art of 1966 in New York City, his striking design of the three inverted steps produces a unique iconic cantilevered building. In the tower portion of the Armstrong Rubber Co. Headquarters in West Haven, a gigantic symmetrical cantilever is produced by hanging floors from the cantilevered rooftop truss structures. This cantilever creates a grand opening between the podium and tower portions of the building. Significant cantilevers in buildings are discussed in much more detail in Chapter 3.

Another variation of tubular steel cantilever chairs was designed by Giuseppe Terragni. In Terragni's cantilever armchair, the continuous front leg and seat frame is cantilevered from one end of the base frame and the backrest frame is vertically cantilevered from another end of the base frame. Since the backrest frame is vertically cantilevered directly from the base frame, the overall configuration of Terragni's cantilever chair looks similar to the non-cantilevered traditional chair with front and back legs. However, unlike the traditional chair, the gap between the horizontal seat cantilever and the vertical backrest cantilever independent but originating from the same base frame is unique. The seat and the backrest cantilevers are designed to produce independently springy behavior. A similarly configured precedent can be found in the Cobra Chair of 1902 designed by Carlo Bugatti, an Italian furniture designer in the Art Nouveau era, though the Cobra Chair is not as springy as Terragni's due to its construction with relatively heavy wood.

Figure 1-8. The Whitney Museum of American Art by Marcel Breuer, New York City, 1966.

Figure 1-9. Armstrong Rubber Co. Headquarters by Marcel Breuer, West Haven, Connecticut, 1970.

Figure 1-10.
Sant'elia Armchair by Giuseppe Terragni.
With permission of Zanotta.

Cantilever chairs have been produced with not only tubular steel but also laminated wood. Alvar Aalto, influenced by tubular steel cantilever chairs, designed a laminated wood version cantilever chair. The basic prototype of the tubular steel cantilever chair frame is a closed loop type structure as shown in Figure 1-2. The seat and the backrest are installed to the frame to complete the chair. In Aalto's cantilever chair, two U-shaped laminated wood frames are placed in parallel to provide the base, front legs and support for the seat. The seat and backrest, constructed with molded plywood as a single L-shaped curved piece, is attached to the two U-shaped frames to complete the chair. While the top and bottom cross members connecting the two side frames are necessary in the typical tubular steel cantilever chairs with seats and backrests made of flexible materials for lateral stability, these are not

Figure 1-11. Cantilever chair by Alvar Aalto. With permission of jacksons.se.

Figure 1-12.
Reverse cantilever chair with removed front legs.
With permission of Titan Furniture.

necessary in Alto's cantilever chair because the solid L-shaped curved plywood rigidly connected with the two parallel U-shaped frames provides the required lateral stability.

Figure 1-12 shows another type of cantilever chair. In this chair, the front legs are removed and the rear legs are extended horizontally to the base location of the removed front legs. Figure 1-13 shows the bending moment

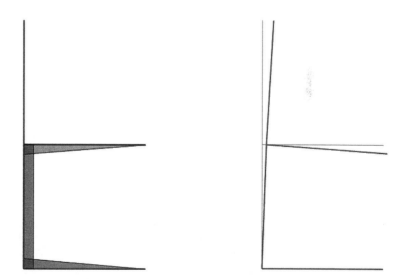

Figure 1-13. Bending moment diagram and deformed shape of the reverse cantilever chair shown in Figure 1-12.

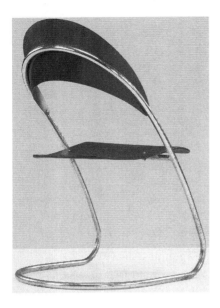

Figure 1-14.
ST14 of 1931 by Hans and Wassili Luckhardt.
With permission of Wright.

and deformed shape of this chair at its simplified form subjected to the gravity load on the seat. The bending moment diagram is basically the same as that of Mart Stam's cantilever chair. However, this reverse cantilever configuration produces deformation downwards as well as towards the front. This is probably not the best direction of displacement in general for people who use chairs to seek comfortable relaxation.

Hans and Wassili Luckhardt's cantilever chair, ST14, of 1931 is, in a sense, radically different from the initial prototype by Mart Stam. In the prototype by Stam, the first rounded 90 degree turn from the base frame produces the front legs, and the second and third turns provide the supports for the seat and backrest, respectively. In Luckhardt's version, only one rounded acute angle turn is made from the base frame to provide the support for the molded plywood backrest at the end. A very thin molded plywood seat is placed between the base and the backrest on the additional metal frame members cantilevered from the main tubular steel frame. In terms of the gravity loads applied to the seat, the performance of Luckhardt's cantilever chair is similar to that of Mart Stam's cantilever chair.

Gerrit Rietveld's Zigzag Chair of 1934 shown in Figure 1-15 is also a strikingly different looking cantilever chair, the leg of which is diagonally arranged. The zigzag form was not new in chair design. The Sitzgeiststuhl by Heinz and Bodo Rasch shown also in Figure 1-15 already employed a zigzag form in 1927. Figure 1-16 shows the bending moment diagram and deformed shape of the Zigzag Chair regarding gravity load applied to the seat. The diagonal arrangement produces moment reversal along the leg, and the overall deformation of this chair, which is straight downward, is relatively small.

Figure 1-15. Zigzag Chair by Gerrit Rietveld (left) and Sitzgeiststuhl by Heinz and Bodo Rasch. With permission of 1stdibs (L), Elsa Mickelsen @ miniaturechairs.com (R).

Less deformation is almost always desirable in buildings. For chairs, however, appropriate level of deformation may increase the user's experience of springy comfort. Not only the amount of deformation but also the direction of deformation is an important factor which influences the user's comfort. Stam's and Mies' cantilever chairs deform downwards as well as towards the back. While this combination of deformations is typically most desirable for chairs, Rietveld's chair deforms almost only downwards when carrying the gravity loads applied to the seat.

Figure 1-16. Bending moment diagram and deformed shape of the Zigzag Chair by Gerrit Rietveld.

The structural analysis of the Zigzag Chair was performed based on moment connections between the straight members. In fact, one of the most challenging aspects of making the Zigzag Chair is creating moment connections between the members because the chair is made of wood panels. Unlike steel, it is very difficult to make moment connections between wood members, especially when the designed form must be kept without adding bracings. In Rietvelt's Zigzag Chair, wood wedges and metal screws were used to make moment connections between the base and the leg and the leg and the seat. Dovetailed joints and metal screws were used between the seat and the backrest. With these elaborate construction methods, the form of the final product is deceivingly simple.

Cantilever chairs presented thus far are produced by assembling multiple pieces made of one or more materials. In cantilever chairs using continuous tubular steel frames, seats and backrests of different materials, such as fabric, leather, or wicker, are installed to the frames to complete the chairs. In Aalto's cantilever chair, only wood is primarily used. But, the chair is produced by assembling two identical U-shaped frame pieces and a curved panel piece for the combined seat and backrest. In Rietveld's Zigzag Chair, the chair is also primarily made of only wood. The design looks as if the chair were produced by folding a rectangular wood panel. However, in reality, the chair is produced by four pieces of wood panels connected elaborately to produce moment connections between them.

Verner Panton was fascinated by the idea of producing a cantilever chair of a single piece made of a single material. In his S-chair of 1956, he initially accomplished this goal by producing an S-shaped chair by curbing a single piece of laminated wood panel. This is in a sense a single piece version Zigzag Chair with curved joints. Soon after, he further developed this idea using plastic. In 1960, he designed the first single piece plastic cantilever chair, the Panton Chair. Though it looks very simple, it took several years to mass-produce the first version of the Panton Chair because of the technical difficulties of actually producing this chair of desired strength and flexibility with a piece of plastic thin enough for stacking. After the first version in rigid polyurethane foam, it had to go through several revisions in terms of design, manufacturing process and material. Since 1999, the injection molded polypropylene version has been mass-produced.

In terms of configuration, the Panton Chair combines the characteristics of Stam's cantilever chair and Rietveld's Zigzag Chair. Let's imaginarily separate the single piece Panton Chair into its continuous loop type curvilinear edges and curved planar infill. The form produced by the curvilinear edges of the Panton Chair looks like a smoothly curved version frame of Mart Stam's cantilever chair. The backrest and the seat are created by infilling the space surrounded by the corresponding curvilinear edges with curved plane. The infilling strategy changes for the front leg and the base. Infilling space between the front leg and base defining edges is done integrally, which creates a

Figure 1-17.
S Chair (left) and Panton
Chair (right) by Verner
Panton. With permission
of 1stdibs.

diagonally curved plane under the seat. This produces more comfortable leg-room, a stable structure and an elegant form.

Chairs are often subjected to lateral loads, especially towards the back. A user often leans against the back of the chair for enhanced comfort. Then, this creates lateral loads and consequently additional bending moments. In Stam's and Mies' cantilever chairs, these lateral load-induced bending moments are added to the gravity load-induced bending moments. Therefore, the structures of the chairs are more stressed and deformed. Figure 1-18 shows bending moment diagrams of the two cantilever chairs subjected to the combined gravity and lateral loads. An important issue to be considered in these cases is that as the lateral load becomes larger, the chairs become vulnerable to overturning failure. Increasing the gravity loads, lowering the seat (and consequently the height of the chair) and increasing horizontal extension of the base frame all help reduce the overturning failure tendency.

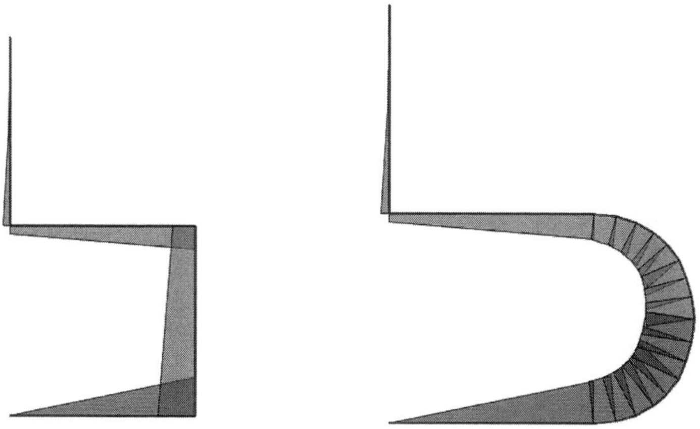

Figure 1-18.
Bending moment diagrams
of cantilever chairs by Mart
Stam (left) and Mies van der
Rohe (right) subjected to
gravity loads on the seat
and lateral loads on the
backrest.

Figure 1-19.
Bending moment diagrams of the Zigzag Chair by Gerrit Rietveld (left) and the reverse cantilever chair shown in Figure 1-12 (right) subjected to gravity loads on the seat and lateral loads on the backrest.

The same concept can be applied to buildings because any building can be considered as a vertical cantilever against lateral loads, such as wind or seismic loads. These loads exert overturning moments to the building. The self-weight of the building produces the counteracting moment. If the counteracting moment is smaller than the overturning moment, the building is vulnerable to overturning failure. Shorter and wider buildings are less vulnerable to overturning failure because shorter buildings are subjected to less overturning moments and wider buildings produce greater resisting moment. The issue of overturning moments and how to efficiently resist them in buildings subjected to lateral loads will be discussed in more detail in Part II.

When Rietveld's Zigzag Chair and the reverse cantilever chair shown in Figure 1-12 are subjected to lateral loads, the gravity-induced bending moments of some framing members are reduced because the lateral load-induced bending moments counteract them. Figure 1-19 shows bending moment diagrams of these chairs subjected to gravity loads on the seats and lateral loads on the backrests.

1.2 CANTILEVER TABLES

Cantilevers are often used to create more sculptural aesthetics or to better produce desired performance. In Eileen Grey's bedside table of 1927, cantilever is employed to satisfy the intended functional performance with its unique aesthetic expression. The circular table top and base frames are connected by vertical legs placed at one end. The resulting C shaped cantilever table functionally serves very well for the user on the bed. Structural behavior of the C shaped cantilever table is similar to that of Mart Stam's previously presented cantilever chair in terms of carrying gravity loads. The opening at

Figure 1-20.
Eileen Grey Table.
With permission of 1stdibs.

the circular base frame is to provide room for a bed leg to pass through when the table base is inserted under the bed.

The Arabesco Table of 1949 by Carlo Mollino for the living room of Casa Orenga shown in Figure 1-21 has a unique form for a piece of furniture belonging to the mid-20th century. It is composed of curved perforated plywood frames and two layers of glass panes. Both the plywood frames and

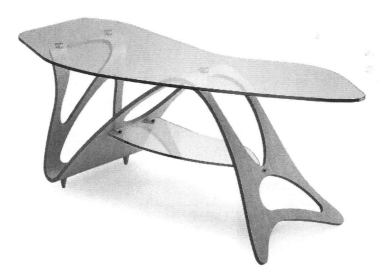

Figure 1-21. Arabesco Table by Carlo Mollino. With permission of Zanotta.

glass panes have irregular free form, within which one can also find structurally sound configuration and efficient proportioning. The curved plywood frame in conjunction with the glass panes is configured in triangular forms, which produce an excellent structural performance. The cantilevered length of the table top glass pane is about 35 percent of the main back span. A cantilever with this proportion performs better structurally in terms of strength and produces less deformation than the case with no cantilever. Performances of cantilevers of various different proportions are discussed in much more detail in Chapter 3.

Bookshelves for dormitory rooms of the Tunisian University in Paris shown in Figure 1-22 were designed in 1952 by Charlotte Perriand in collaboration with Jean Prouve and Sonia Delaunay. While the asymmetrical composition produces a very dynamic expression, the strategic massing about the two main vertical supports on the floor gives a strong impression of structural stability. Though, in fact, the bookshelf was additionally supported by wall mountings, even just with the two vertical supports, it could be supported safely. For the longest main horizontal member, one vertical support is positioned around the center and the other support, towards only one end of the member. Without the strategically placed four layers of shelves, the structure would be very vulnerable to overturning failure even with slightly larger loads applied to the longer cantilever side of the main horizontal member because the end support away from the longer cantilever would be lifted up. However, with the help of the weight provided by the stacked shelves, the end support will not be lifted up and the stability of the entire structure can be obtained. This type of asymmetric configuration and strategic massing to prevent overturning failure can often be found in cantilevered buildings as will be discussed in much more detail in Chapter 3.

Figure 1-22. Bibliotheque pour la maison de la Tunisie by Charlotte Perriand. With permission of Wright.

Figure 1-23. Cantilever Table by Rainer Spehl. With permission of Rainer Spehl.

The cantilevered table designed by Rainer Spehl shown in Figure 1-23 has a similar C shape configuration to Eileen Grey's table. While the length of the base of Eileen Grey's table is similar to the cantilevered length of the table top, the length of the base of Rainer Spehl's table is much shorter than the cantilevered length of the table top. Structurally, as the length of the base becomes shorter, the table becomes more vulnerable to overturning failure based on gravity loads applied to the table top. Overturning would occur about the free end of the base. In terms of the applied loads, the overturning moment is increased as the applied load placed on the portion of the table top closer to the free end is increased. On the contrary, the counteracting moment is increased as the applied load placed on the portion of the table top closer to the vertical leg is increased. In addition, any self-weight within the width of the base, including the base itself, the leg and the corresponding portion of the table top, participates in producing the counteracting moment. In the table designed by Spehl, the table top is made of relatively light wood panel and the vertical leg and base are made of much heavier concrete in order to increase the self-weight-induced resisting moment to help prevent overturning failure. The structural issue caused by the very challenging geometric configuration was resolved to a large degree by material choices in this project. And these material choices in conjunction with the diving board-like unusual form of the table produce very dynamic aesthetics.

Balancing the loads in asymmetrically configured cantilevers is important to prevent gravity induced overturning failure. In buildings, overturning tendency of the asymmetrically configured cantilever structures can be produced by not only gravity but also lateral loads. The impact of the combined gravity and lateral loads could be significantly larger than the individual loads. However, the overturning failure of buildings can be prevented by the foundation system typically hidden under the ground. The geometric configuration of the foundation system can be determined in such a way that the overturning failure can be better prevented, or deep foundations with tensile capacity typically by surface frictions can be employed. Some of the cantilever buildings presented in Chapter 3 include these cases. In cantilevered tables which are generally not anchored to the floor, however, the overturning failure should be prevented by the table structure itself configured to balance the loads.

The SMT coffee table designed by J. Wade Beam is another C-shaped cantilevered table. In this table, a 1/2 in. thick tempered glass top of square shape is diagonally cantilevered from the tilted 1 in. thick stainless steel base. If the stainless steel base of the same size had been deigned vertically at the end of the table to support the cantilever, its width would not have been large enough to prevent gravity-induced overturning failure. The stainless steel base is tilted towards the opposite direction of the glass top cantilever. Therefore, the table is eventually configured with the tilted shorter cantilever of heavier steel and the longer but lighter glass cantilever projected in the opposite direction to the tilted base cantilever to balance the load. This careful geometric configuration along with the materials choice produces a unique dynamic expression and required structural stability at the same time.

In fact, the name of the table, SMT, reflects the difficulty of balancing the loads when assembling this table according to Deborah Cvirko at Brueton which produces the SMT table. The construction of this table was challenging and it was difficult to achieve a perfect balance. The form of the stainless steel base of the table had to be tilted to balance the loads of the completed

Figure 1-24. The SMT coffee table designed for Brueton by J. Wade Beam.

table. However, the tilted base itself tends to easily fall over until the glass top is installed. Therefore, the engineers were saying the table was "So Much Trouble" during the prototyping and the initials were used to name the SMT table.

In the rectangular table shown in Figure 1-25, the table top is symmetrically cantilevered in both directions perpendicular to each other. In the longitudinal direction, the length of the table is divided into three zones, the central zone and the two end cantilever zones. With the length of the both cantilevers of about 20 percent of the total length, the weight of the table top and distributed loads on it can be carried more efficiently, compared with the case with no cantilevers. Figure 1-26 shows comparative bending moment diagrams of the table top with alternative support locations and connections. The first one is the case when the table top is simply supported at both ends. When the connections between the vertical supports and the table top are rigid to provide lateral stability, the bending moments are reduced as can be seen in the second diagram. When both vertical supports are pushed in by about 20 percent of the total length, the bending moments are further reduced as can be seen in the third diagram. Both symmetrically and asymmetrically cantilevered buildings and optimal proportioning of cantilevers are discussed in greater detail in Chapter 3.

The two transverse direction beams which carry the table top loads have two symmetrical cantilevers about the central vertical supports. This configuration, which produces all negative bending moments throughout the beams, is not the optimal cantilever condition in terms of structural perform-

Figure 1-25. Symmetrically cantilevered rectangular table. Photographer: Hugh Hartshorne, Designer: Stephen Hammer of Urban Forest Furniture.

25

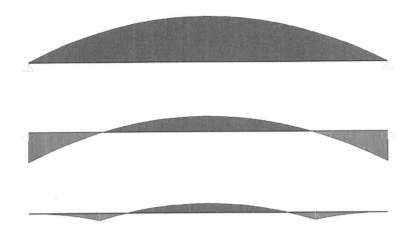

Figure 1-26. Comparative bending moment diagrams of the table top shown in Figure 1-25 with alternative support locations and connections.

ance. However, this configuration makes the otherwise typical perimeter legs moved far away from the edges of the table. In order to prevent the over-turning failure of the table in the transverse direction due to any unbalanced applied loads, the horizontal bases are extended in that direction from the bottom of the central legs. In terms of overall geometry, due to the symmetrical configuration, overturning tendency of the table shown in Figure 1-25 is smaller than that of asymmetrically configured cantilevered tables.

CHAPTER 2
CANTILEVERED BUILDING COMPONENTS

AS A CONSTRUCTED OBJECT, a building is composed of many different physical components. Some small components participate as parts of larger components, which become even larger systems, and this process eventually results in the completion of an entire building. This chapter presents structural principles and related design issues of cantilevered building components, such as cantilevered stairs, balconies and canopies, theoretically and with real world examples.

2.1. CANTILEVERED STAIRS

Stairs are an essential building component which connects spaces of different levels. This unique architectural function requires careful structural considerations and at the same time provides a good opportunity to explore creative design. Therefore, it is not uncommon that the state of building technology and architectural design trend of a specific time period can be observed through the design and construction of stairs. Structural and architectural design integration for stairs often produces very innovative solutions.

Stairs of many different configurations are designed depending on the project specific situations. Cantilevered stairs have been used in buildings since ancient times. They are often designed to resolve certain design problems which may be difficult to solve with non-cantilevered stairs and usually perform as very dramatic architectural design elements. This section presents cantilevered stairs of various configurations and materials.

2.1.1. Early Cantilevered Stone Stairs

Stone is one of the oldest building construction materials. While stones used as building materials, such as granite, limestone and marble, have great compressive strength, no stone has substantial tensile strength. Therefore, stone is typically not a good material for cantilever structures which carry loads primarily by bending action because bending is a combined action of compression and tension. Nevertheless, cantilevered stone staircases can be found from buildings of the Hellenic period of 500–300 BC. Cantilevered stairs reappeared in the Islamic world in the medieval period and were reinvented in Western Europe during the Renaissance. Many cantilevered stone stairs were helical and the cantilevered stone block treads were supported within the thickness of the internal walls.

A dramatic example of very old cantilevered stone stairs in the Hellenic period can be found in the remains of the cantilevered staircase in the tower of Agios Petros on the island of Andros as can be seen in Figure 2-1. The cantilevered stairs at that time were built with local schist. About two millennia later, cantilevered stone staircases were revived in Andrea Palladio's buildings in the 16th century, such as the Convento della Carita in Venice. Influenced by Palladio, Inigo Jones designed the cantilevered stone Tulip Staircase, which employed the rebate for the first time, in the Queen's House of 1635 at Greenwich. During the 17th through 19th centuries, cantilevered stone staircases were widely employed in the buildings of the UK.

Figure 2-1. Remains of the cantilevered stone staircase in the tower of Agios Petros on the island of Andros. With permission of Michael Tutton.

Figure 2-2. Tulip Staircase in the Queen's House at Greenwich.

Another excellent UK example with rebates can be found in Somerset House in London designed by Sir William Chambers in 1795. In typical cantilevered stone stairs with rebates, cantilevered stone block treads are embedded into the supporting wall by about 10 cm. The stone block treads above and below slightly overlap each other with rebates which are interlocking connections between the treads. Based on the embedment at one end and overlapping of the treads, the self-weight of the treads and applied loads on them are not carried by true cantilever action but by compression and torsion. The first tread on the floor is usually completely supported by the floor with no torsion. Therefore, the second tread just above the first one on the floor develops the largest torsion. And the last tread just below the landing carries the smallest loads and develops the smallest torsion. When the stone treads are constructed with rebates, additional torsional resistance is provided by the rebates compared with the case with no rebates.

Cantilevered stone stairs constructed by experienced master masons usually survive hundreds of years. However, some old worn-out cantilevered stone stairs collapsed or were found to be broken due to poor management. The cantilevered stone stairs usually fail around the support area, where the torsional stress is larger as can be seen in Figure 2-5.

Figure 2-3. Stamp office staircase at Somerset House. With permission of Russell Taylor Architects.

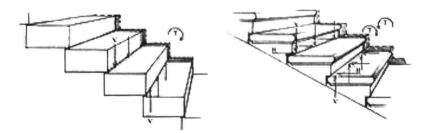

Figure 2-4. Cantilevered stone staircase without and with rebates. With permission of Russell Taylor Architects.

Figure 2-5. Cracking of stone cantilevered stairs. With permission of Helen Rogers www.stonestairs.net.

2.1.2. Cantilevered Stairs of Modern Materials

Stone is not appropriate material to build true cantilevered stairs because of its unreliable tensile capacity. Overlapping the treads was a good strategy to construct the previously presented cantilevered stone stairs in the 17th through 19th centuries. The limitation of the material was partially overcome by overlapping the stone treads. However, the loads of and on the treads were not carried by true cantilever action because of the overlapping, and their expressions of cantilever were limited. Today, with more appropriate modern structural materials, true cantilevered stairs can be constructed with no overlapping between the treads.

Figure 2-6 shows an example of reinforced concrete cantilevered stairs in the Hanasaki House in Yokohama designed by MoNo. The textural characteristic of exposed concrete is similar to that of stone compared with other building materials. Unlike traditional stone cantilevered stairs, the treads of this staircase are constructed with no overlapping. They are independently cantilevered from the reinforced concrete vertical supporting wall. Due to the unique characteristics of cast-in-place reinforced concrete structures, the

31

Figure 2-6. Hanasaki House in Yokohama by MoNo. With permission of MoNo.

vertical supporting wall and independent treads are monolithically connected in this staircase, and rigid connections, necessary for the cantilever action, are produced between them with appropriate reinforcements. The bending mechanism caused by vertical loads applied to the independent treads produces internal tensile forces towards the top surfaces and compressive forces towards the bottom surfaces of the treads. Since concrete does not have substantial tensile capacity, steel reinforcing bars are placed towards the top surfaces of the treads during the construction process before pouring concrete to carry the tensile forces. The compressive forces towards the bottom surfaces are resisted by concrete there. Railings are typically required for stairs by building codes for safety. In the Hanasaki House, removable railings were installed.

Steel is a good material for cantilever structures because it can excellently carry both tensile and compressive forces. In the steel cantilever stairs designed by Lawrence Architecture for a residence in West Seattle, steel treads are cantilevered from reinforced concrete walls. In order to make rigid connections between the reinforced concrete wall and the cantilevered steel treads, steel plates are typically embedded into the reinforced concrete wall. In general, shear studs are welded to the back side of the steel plates and embedded into the concrete wall together with the plates. Once the formwork of the reinforced concrete wall is removed, the embedded steel plates are exposed. Then, the steel tread structures are welded to the steel plates. This is a typical method to make moment connections between wall type reinforced concrete structures and beam type steel structural members.

The Kaze No Oka Crematorium designed by Fumihiko Maki also includes a similar cantilever staircase composed of reinforced concrete wall,

Figure 2-7. Steel cantilevered stairs. Photographer: Benjamin Benschneider, Lawrence Architecture.

steel tread structures cantilevered from the wall and wood finish for the treads. In this staircase, the reinforced concrete wall supporting the cantilevered treads is a sloped deep beam type structure, the two ends of which are supported at different levels. In conjunction with the triangular void space under the sloped beam type supporting wall, the hovering expression of the cantilevered treads is much emphasized. The height of the sloped reinforced concrete wall terminates in such a way that it can also perform as a handrail of the staircase. The railing on the free end side of the cantilevered treads is composed of very thin and light steel members to minimize the load on the cantilever and emphasize its expression at the same time.

The cantilever staircase shown in Figure 2-8 is employed for a light wood frame house. Compared to steel or reinforced concrete, wood is generally less strong, more flexible and less stable material. In addition and more importantly, making moment connections with wood is much more difficult, compared with making them with steel or reinforced concrete. Therefore, steel is often combined in wood structure buildings to resolve some challenging structural issues. In the cantilevered stairs shown in Figure 2-8, a steel stringer is embedded into the typical light wood frame stud wall, and the steel tread and riser structure is cantilevered from the embedded steel

Figure 2-8. Cantilever stairs in a light wood frame house. Architect: Anthony J. Ries, Structural Engineer: Jim Houlette.

stringer using moment connection by welding between them. Alternatively, independent tread structures without risers can be welded to the stringer to produce more dramatic hovering expression of the cantilevered stairs. The steel stringer embedded into the wall and steel tread structures can be fully covered by wood finishes so the design of the stairs can be coherent with other architectural design features of the light wood frame house when desired.

In cantilevered tread structures, the maximum bending moment is developed at the support and the minimum at the free end. Following this structural logic, the treads of cantilevered stairs can be designed to have greater structural depth towards the support unlike the prismatic form treads presented so far. Figure 2-9 shows tapered tread structures with the greatest structural depth at the support to take the maximum bending moment there.

Glass is sometimes structurally used to produce dramatic cantilever stairs. Though normal float glass cannot be used directly for stairs, by appropriate heat treatment, glass treads of very high strength can be produced. The helical staircase at a private house in Scotland shown in Figure 2-10 uses cantilevered treads made of laminated heat soaked toughened glass 40 mm thick including structural sentry interlayer. Toughened glass (or tempered glass) is generated by heating float glass to a temperature over 600°C and cooling rapidly over a period of 2–10 seconds. By this process, the original float glass already exactly cut to fit specific needs becomes stronger and can take greater bending moment because the pre-compressed surface of the toughened glass is able to take otherwise large tensile stresses developed by bending. Heat soaking is a process to reduce the risk of abrupt failure of

Figure 2-9. Cantilever stairs with tapered tread structures coated with travertine marble type resin (left) and made with Corten steel plates (right). With permission of Marretti USA.

toughened glass due to nickel sulfide inclusions. By heating the toughened glass to 290 degrees for a given period of time and slowly cooling it before use, it is likely to shatter if it contains nickel sulfide.

An 80 mm wide helical steel stringer is used to support the glass treads which is cantilevered 1300 mm. The stringer is manufactured to have rectangular pockets into which the glass treads are inserted. The glass treads are secured by a clamping system built in the pockets and high modulus silicon. With this configuration, each tread of laminated heat soaked toughened glass can take up to 1 metric ton point load at the free end of the cantilever according to the manufacturer. The stairs are built around a cylindrical glass aquarium with a gap of 40 mm between the tips of the cantilevered treads and the exterior glass surface of the aquarium. This composition with the small gaps close to the free ends further enhances the already dramatic expression of the cantilevered glass stairs.

In the cantilevered stairs presented thus far, one end of the treads is rigidly connected to the vertical support structures and the other ends are free from any type of support. In addition to this common configuration, the concept of cantilever can be used for stairs with different configurations. Figure 2-11 shows a floating staircase in the Coach House in Wimbledon. As can be seen in the figure, the entire stairway including the intermediate landing is cantilevered. This design strategy produces a cantilevered staircase of truss-like triangular geometric configuration, which is structurally very efficient to carry applied loads.

In the Itamaraty Palace in Brasilia, Oscar Niemeyer designed a similarly cantilevered staircase in a more sculptural helical form as can be seen in Figure 2-12. The reinforced concrete stringer for the stairway without any intermediate

Figure 2-10. Glass cantilever stairs at a private house in Scotland. With permission of Julian Hunter Architects.

Figure 2-11. Floating staircase in Coach House. With permission of Hale Brown Architects, demax.co.uk.

landing is spirally cantilevered with one end supported by the ground level and the other by the second level. The reinforced concrete treads are cantilevered again symmetrically from the central helical stringer. The central stringer is relatively wide, and the right angle cuts of the central stringer to hold treads provide risers. However, the cantilevered portions of the treads beyond the central stringer do not have risers. This configuration emphasizes the floating image of the stairway. A runner is placed over the central portion of the treads directly supported by the stringer, and the cantilevered portions of the reinforced concrete treads are exposed without being covered by the runner. Since this stairway does not have railings, the runner placed at a certain distance from the tips of the cantilevered treads visually guides the users for safety. Integrated with the continued spiral staircase from the ground to the basement floor, the cantilevered spiral stairway shown in Figure 2-12 provides a unique architectural experience to the users.

In the Saitama Prefectural University designed by Reiken Yamamoto and Field Shop, a series of multiple story outdoor stairs are cantilevered beyond the exterior walls. The uppermost floor stairs are supported by the truss structure cantilevered from the building and integrated with the stairway's flights and landing. The sloped truss members along the upper and lower flights work in tension and compression, respectively. The tension and compression members meet with pin type connections under the landing. The stairs below the uppermost one are hung from the truss structure by tension rods.

Figure 2-12. Helical staircase in the Itamaraty Palace in Brasilia. With permission of Adam Gebrian.

Figure 2-13. Staircase in National Archives of France. Photographer: Kamal KhalfiStudio, Studio Fuksas.

In the National Archives of France, the staircase shown in Figure 2-13 is cantilevered in a different orientation. In this example, the intermediate landings are cantilevered from the wall and the flights span between the cantilevered landings. In this configuration, the entire flights may be designed to be detached from the wall from which the landings are cantilevered to emphasize the visual expression of floating.

Beginning from the ancient stone cantilever stairs, the history of cantilever stairs is very long. The configuration of the old stone cantilever stairs was much limited due to the property of stone which lacks tensile capacity. Today, however, cantilever stairs of more innovative configurations are designed and constructed with modern structural materials which have greater bending resistance, such as reinforced concrete and steel. Recently, even laminated heat-treated glass is not uncommonly used for stairs, including cantilevered stairs, the design of which, especially with glass, is usually very challenging.

2.2. CANTILEVERED BALCONIES

A balcony is typically a projected platform beyond the exterior wall of a building usually protected by railing. Some balconies are completely enclosed by glass window walls and roof. According to Viollet-le-Duc, the history of the external cantilevered balcony dates back to the 11th century "hourd" – a temporary wood structure for battles. Figure 2-14 shows a restored hourdage by Viollet-le-Duc at the Cite de Carcassonne. The temporary hourd began to disappear and was replaced by permanent stone battlements from the 14th century.

Another origin of exterior cantilevered balconies can be found in mashrabiya which is deeply related to traditional Arabic culture and climatic condition. The mashrabiya is typically a small cantilevered oriel window with limited openings. Through the grill or louver-like openings, natural ventilation is possible. However, direct visual connections between the interior and exterior are prohibited so women in the house cannot have direct contact with the outside world. The history of mashrabiya dates back to the 7th century.

With these historical backgrounds, cantilevered balconies began to prevail as an important architectural element from the 17th century in Malta. Maltese cantilevered open stone balconies were supported by brackets called saljaturi which were typically very ornamental. The brackets are supported by load bearing walls and are thickest where they begin from the load bearing wall and become thinner as they reach the end. The length of the cantilever supported by stone brackets is limited because stone has minimal tensile capacity. From the mid-18th century, closed wooden balconies began to gain popularity. Wooden balconies were typically painted with bright color oil paints to protect the material from weather. Stone brackets were still used as structural supports for the wooden balconies in many cases.

Figure 2-14. Restored hourd in Carcassonne.

Balconies of modern buildings are structured with modern building materials such as reinforced concrete and steel. Balconies are also constructed with wood for light wood frame structures, which are prevalently used for residential buildings in the US. Three typical strategies to support cantilevered balconies are using brackets underneath the balconies, hanging the balconies from above using sloped tensile members, and directly cantilevering floor structures, such as wood floor joists, steel floor beams or reinforced concrete slabs.

Figure 2-15. Stone corbels known as saljaturi in Maltese balconies.

Figure 2-16. Maltese closed wooden balconies.

Using brackets and hangers requires additional structural members to support balconies. The horizontal floor supporting members of the balconies and the additional structural members such as hangers create triangular geometric configurations along with the vertical supports which are typically an integral part of the main building structure. By this triangulation, the load carrying mechanism of balconies is primarily done by axial actions, which is usually very efficient structurally. Certainly, bending moment is also developed in the horizontal floor members of the balcony with applied vertical loads on them. For the triangular configurations, no moment connections are necessary, though moment connections are often necessary for the railing system of the balcony for safety.

When using sloped hangers or brackets, the angles created between the horizontal floor support members of the balcony and the hangers or brackets play an important role structurally and aesthetically. As the angles become larger, the axial forces developed in the sloped members become smaller, and consequently, the member sizes can be smaller. On the contrary, as the angles become smaller, the axial forces become larger, and the member sizes should be larger. This structural logic should be carefully integrated with architectural design.

While cantilevered balconies supported by hangers and brackets require additional structural members to support them, directly cantilevering

Figure 2-17. Steel balcony supported by steel hangers. With permission of InnoTech Manufacturing, LLC.

Figure 2-18. Steel balcony supported by steel brackets.

interior floor structures to make balconies does not require additional hangers or brackets, and produces simpler visual expressions. Steel, reinforced concrete and wood are all feasible materials to design and construct directly cantilevered balconies which carry applied loads primarily by bending actions. In Bauhaus shown in Figure 2-19, reinforced concrete balconies are produced by directly cantilevering interior floor slabs beyond the façade plane. The simple railing is integral with the cantilevered floor to complete the balcony.

By cantilevering interior floor structures beyond their perimeter supporting members with a certain structurally desirable proportion of the interior back span to external cantilever, structural efficiency of the flooring system may even be enhanced. (See Figure 3-7 in Chapter 3.) However, projecting interior floor structures passing through the façade plane creates systematic thermal bridges between the conditioned indoor and unconditioned outdoor environments. The problem of thermal bridging is more critical in steel and reinforced concrete structures than in wood. Among these three structural materials, steel has the highest thermal conductivity, while wood has the lowest. In order to resolve the issue of thermal bridges through cantilevered balconies, structural thermal break systems have been developed by some manufacturers and are readily available for easier construction. Insulation is placed between the interior and exterior portion of the continuous cantilevered structure. This location of insulation should be carefully determined so that it is on the same plane with the typical exterior wall insulation.

Figure 2-19. Reinforced concrete balconies at Bauhaus. With permission of Yvonne Tenschert, 2009, Bauhaus Dessau Foundation.

In the structural thermal break system employed for steel structures, steel beams on each side of the insulation are connected with fasteners which penetrate the insulation. In reinforced concrete structures, both tension and compression reinforcements continuously pass through the insulation. Since the steel beam connecting fasteners and reinforcing bars pass through the insulation, a certain level of systematic thermal bridging action still exists in the structural thermal break systems. However, the thermal conductivities through

Figure 2-20.
Steel structural thermal break system. With permission of Fabreeka International, Inc.

Figure 2-21.
Reinforced concrete structural thermal break system. With permission of Ancon Building Products.

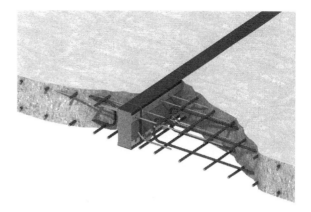

the structural thermal break systems are much smaller than those through the normal structures without the systems.

The Haus mit Veranden in Vienna designed by architects Rüdiger Lainer is an apartment complex with 254 dwelling units. Every unit has outdoor spaces, and many of them are large cantilevered balconies. These either cantilevered or Juliet balconies naturally connect each unit with the outdoor environment. The dramatic projection of large cantilevered balconies adds sculptural quality to the complex composed of irregular form multistory building masses which already have highly sculptural quality. Since the balconies are structured by cantilevering interior floor structures, they can work

Figure 2-22. Haus mit Veranden. Photographer: Hubert Dimko, Architect: RLP Rüdiger Lainer + Partner.

45

as systematic thermal bridges. A thermal break system similar to what is shown in Figure 2-21 was employed to resolve this issue and save energy for environmental control.

VM House, Copenhagen, Denmark

The VM House in Copenhagen, Denmark, designed by JDS Architects is composed of two residential buildings facing each other. The plans of the two buildings were designed to have the forms of letters V and M to provide the residents with better views. There are 230 apartment units ranging from single houses to family houses. In order to provide unique living environments for the residents there are 120 different floor plan types in the complex.

The V-House is characterized by sharp triangular shape cantilevered balconies. The shape and arrangement of the pointed balconies were determined to allow maximum daylight for each unit and at the same time to provide space for communication between neighbors. In order to better support the relatively long triangular balconies, a point close to the outer tip of the triangle is connected to the primary structural member of the building with a steel rod. Typically two tensile members are required for conventional rectangular hung balconies. In the V-House, however, only one tensile rod is fine. Due to the triangular shape of the balcony, a three-dimensional truss in the form of a tetrahedron is produced with only one tensile rod. The carefully configured triangular cantilevered balconies have added an unprecedented unique expression to the building and integratively perform well.

Figure 2-23. VM House balconies. Julien de Smedt Architects

The Ledge at Willis Tower, Chicago, USA

The Ledge at Willis Tower (formerly known as Sears Tower) in Chicago is composed of four identical closed balconies unique in many aspects. Located on the 103rd floor of Willis Tower, 412 m from the ground, the Ledge's cantilevered portion is mostly constructed with glass panels to provide spectacular views of Chicago. In general, balconies are cantilevered to provide a view outward. In the Ledge, it was important to provide a view not only outward but also downward through the transparent glass floor.

The Willis Tower of 1973 did not have the Ledge until 2009. As a later addition to the building, the rectangular box form Ledge was structured with steel frames and glass panels. It can be cantilevered and retracted using the railing system supported by the floor framing of the tower. The retraction was necessary in order not to obstruct the operation of the existing window washing system when in use, and for the maintenance of the Ledge itself. The half of the Ledge structure which always remains inside the building, is composed of steel braced frames. The other half of the structure, which is usually in its cantilevered position, is composed of mostly transparent glass panels. The length of the cantilever is about 1.3 m, and the width and height of the cantilevered balcony is about 3.2 m and 3.6 m, respectively.

In order to support the cantilevered portion when the Ledge is in its usual cantilevered position, the steel members on the both edges of the ceiling are cantilevered and the glass walls are hung from the cantilevered steel members. The glass roof panel of the cantilevered portion is directly supported by the cantilevered steel frame members, and the glass floor is supported by the hung glass walls. Point-fixings are used for the connections between the glass panels in order to maximize the view with minimal obstructions.

For safety purposes, the glass enclosure of the Ledge is made with laminated glass composed of three layers of 12 mm tempered low iron, heat-

Figure 2-24. The Ledge at Willis Tower. With permission of John Kooymans.

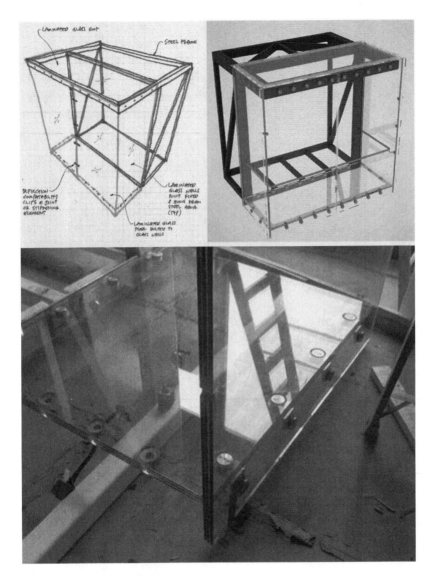

Figure 2-25. Structural details of the Ledge at Willis Tower. With permission of John Kooymans.

soaked glass. In addition, 6 mm sacrificial lite is added on the laminated glass floor to protect it. Safety films were applied to the wall glass panels to protect them from scratching. Low iron glass is used to provide the view of superior clarity from the Ledge. By reducing iron content, which produces green tint in regular float glass, glass can be truly clear. Since green tint becomes more visible as the thickness of glass is increased and color distortion occurs as regular float glass is laminated, application of low iron glass for laminated glass composed of thick layers can provide dramatic differences in terms of clarity.

2.3. CANTILEVERED CANOPIES

The history of door canopies and awnings goes back to ancient times. While they are installed for their functional performances to protect the entrance doors and windows from the heat of the sun and downpour of rain, their aesthetic contribution to the buildings is also significant. Many contemporary buildings have cantilevered entrance canopies. The structural concept of cantilevered entrance canopies can be very similar to that of cantilevered balconies. Entrance canopies can be very efficiently supported by hangers or brackets by their axial actions or they can be directly cantilevered from the interior floor structures or roof structures.

Figure 2-26 shows an entrance canopy supported by hangers. In terms of carrying gravity loads, the hangers are subjected to tension. However, if strong uplift force due to wind even larger than the gravity load is expected, the hangers should be designed to carry compressive forces as well. The angle between the canopy and the hangers has an impact on the structural performance. The larger, the smaller axial forces in the hangers, and vice versa.

Though the load carrying mechanism of cantilevered balconies and canopies can be very similar, unlike cantilevered balconies, cantilevered

Figure 2-26. Entrance canopy supported by hangers.

Figure 2-27. Entrance canopy of the City Point Building in London.

canopies do not need to provide occupiable platforms in general. Therefore, the design of cantilevered canopies can be more diverse especially in terms of form and material. Canopies are often designed with glass and can also be designed with more flexible materials including fabrics.

Figure 2-27 shows the cantilevered entrance canopy of the City Point Building in London. Following the arch form main entrance of the building, the entrance canopy is composed of a series of tilted steel arches hung from the main entrance arch. And the slanted vaulted form of the entrance canopy created by the series of component arches is covered by glass panels point-fixed to the steel arches.

Figure 2-28 shows the entrance canopy to the Yurakucho underground station in Tokyo. The slanted curved cantilever made primarily of glass makes this canopy sheltering the staircase to the station unique. The cantilevered canopy is supported by three cantilevered beams composed of smaller elements made of laminated glass. The beam elements are connected by stainless steel pins of 40 mm diameter at about their mid-points and end-points with overlapping as shown in the figure. Following the structural logic of the cantilever, the depth of the cantilevered beam elements becomes larger towards the support made of stainless steel round pipe running along the width

Figure 2-28. Entrance canopy to the Yurakucho underground station.

of the canopy. V-shaped stainless steel brackets integrated with the round pipe make connections between the cantilevered glass beams and their pipe support.

Unlike cantilevered balconies which typically require occupiable platforms built with solid and strong structural materials, cantilevered canopies can be designed with flexible materials, such as fabrics, of curved forms. The cantilevered canopy at the East Texas Physicians Alliance made of steel frames and fabric shades the drop-off area and protects the area from the weather. The gently curved longest main cantilever members are supported from the bottom by straight steel members and also hung from the top by steel cables. The angled steel members at the bottom support the main cantilever

Figure 2-29. Cantilever canopy at the East Texas Physicians Alliance in Palestine, Texas. With permission of FabriTec Structures.

members at about their mid-spans and the cables are attached to the points close to the tips of the main members. Due to the resulting triangular forms, the cantilever is supported primarily by axial actions of the component members which are pin-connected. Flexible fabrics are placed over the frames to complete the cantilevered canopy.

Trumpf Campus Gate House, Ditzingen, Germany

The Trumpf Campus Gate House in Ditzingen, Germany, designed by Barkow Leibinger Architects is a small gate house with a dramatic cantilevered canopy. Trumpf is a large manufacturer of fabricating equipment and industrial lasers. It was important for the architect to incorporate the technology of Trumpf in the design and construction of the gate house – the first building of the campus experienced by staff and visitors. Therefore, the design was performed with the Trumpf's laser-cut technology in mind, and the technology was actually used for the construction and finally expressed as constructed form.

The building is composed of the 130 m² enclosed space containing reception, waiting and technical areas. Four columns support the roof of 32 m x 11 m. The roof is cantilevered to all four sides with the longest cantilever of about 20 m in the longitudinal direction over the entrance door and across the street lanes. The cantilevered roof structure in the longitudinal direction is supported by 17 hollow steel box girders. Steel cross beams are placed between the box girders in a zigzag pattern. The box girders are 50 cm deep

Figure 2-30. Entrance canopy of the Trumpf Campus Gate House. Architect: Barkow Leibinger & Photographers: David Franck (T), Corinne Rose (B).

and their width varies from thickest around the columns on the longer cantilever side to thinner around the free ends of the cantilevers. Density of the cross beams of a zigzag pattern also varies following the structural logic. Trumpf's own laser-cut technology was used to produce the structural members of varying sizes and shapes of the roof as shown in Figure 2-30.

53

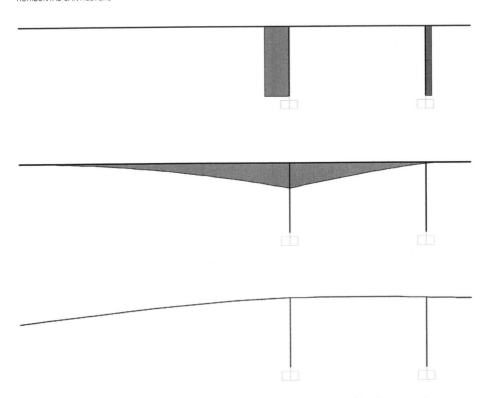

Figure 2-31. Axial force diagram (top, tension in darker shade and compression in lighter shade), bending moment diagram (middle) and deformed shape (bottom) of the simplified structural model of the Trumpf Campus Gate House subjected to uniformly distributed loads on the roof.

The roof structure arrived on site in six pieces and these were bolted together before they were lifted and connected to the four columns. To control large deformation caused by the large cantilever, the roof structure was cambered before installation. In the longitudinal direction, the roof is an asymmetrical two-sided cantilever with an approximate proportion of 1:3:6. With this proportion, the columns on the shorter cantilever side are subjected to tensile forces as can be seen in the axial force diagram of Figure 2-31. This condition makes the structure vulnerable to gravity-induced overturning failure. Therefore, the foundation must be designed to resist the overturning tendency of the entire structure. The thickness of the foundation of the gate house is 90 cm. Without the large cantilever, a foundation thickness of only 15 cm would be required, according to the project engineer. Structural performances of cantilevered buildings based on the cantilever to back span ratios are discussed in greater detail in Chapter 3.

The enclosure system of the gate house is composed of two layers of low-e float glass with a gap of 30 cm between the layers. The gap is filled with acrylic glass tubes, which produce blurry transparency. The double layered enclosure system substantially contributes to energy reduction for air

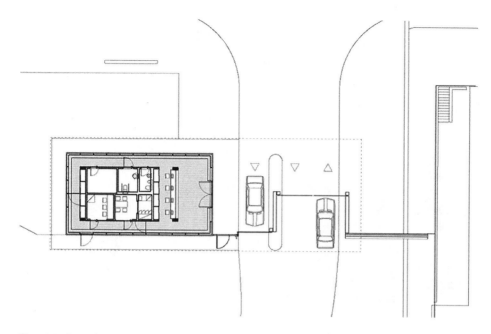

Figure 2-32. Trumpf Campus Gate House plan. With permission of Barkow Leibinger.

conditioning of the building. The columns are set back from the enclosure system which is non-load bearing. Customized rubber gaskets are located between the glass enclosure and the roof structure to safely accommodate the deformation-induced movements of the roof. This configuration and material composition accentuates hovering expression of the large cantilevered roof.

One Central Park, Sydney, Australia

One Central Park in Sydney designed by Ateliers Jean Nouvel is composed of two residential towers of 34 and 16 stories. The two towers are connected by the shared 5-story podium, which has amenity facilities and roof garden. While the complex is configured to maximize natural light, it is still challenging to introduce daylighting into the pocket space between the towers. In order to resolve this issue, a set of heliostats is employed to reflect the sunlight into the deep pocket space including the roof garden and even into the atrium under the roof garden through the skylight.

For the installation of the heliostat, a large cantilever is projected from the taller tower between levels 30 and 32 towards the shorter tower. The heliostat is composed of two sets of mirrors, one on top of the shorter building and the other under the cantilever of the taller tower. The sunlight is reflected from the motorized mirrors on top of the lower tower to the mirrors under the cantilever and finally to the roof garden area between the towers.

Figure 2-33. One Central Park. Design architect: Ateliers Jean Nouvel, Local collaborating architect: PTW Architects.

While cantilevered canopies typically function as sunlight protector, the cantilever structure in One Central Park is primarily used to introduce sunlight into the shaded space of the complex.

In order to support the approximately 40 m long cantilever for the installation of the heliostat system, four sets of two-story tall trusses are employed between levels 30 and 32. The trusses are designed for both the cantilevered portion and the back spans to provide strength and stiffness enough to support the cantilever. The cantilevered trusses function not only as the primary structural support for the heliostat but also as space for the sky garden. Compared to the cantilevered roof of the Gate House of the Trumpf Campus, the cantilever of One Central Park is much longer and heavier. However, the cantilevered trusses are shorter than the back span trusses in One Central Park Tower. Further, applied gravity loads on the cantilevered trusses are smaller than those on the back span trusses, above which there are multiple floors. Therefore, the cantilever structure of One Central Park Tower is not vulnerable to overturning failure due to gravity loads. However, since this is a tall building, lateral load-induced overturning failure must be carefully studied. Lateral load resisting systems for tall buildings are discussed in greater detail in Part II.

Voest Alpine Office Center, Linz, Austria

The Voest Alpine Office Center in Linz, Austria, designed by Dietmar Feichtinger Architectes is a 5-story office building of a gently curved form. The main

Figure 2-34. Voest Alpine Office Center. With permission of Dietmar Feichtinger Architectes, Photo: Josef Pausch.

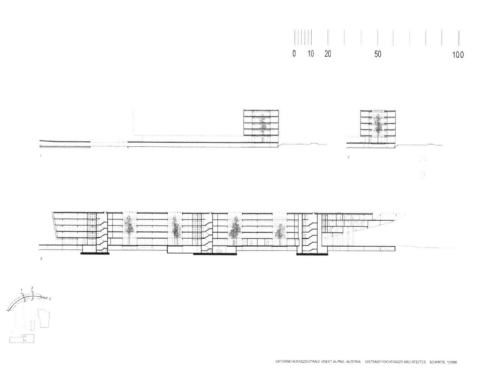

Figure 2-35. Voest Alpine Office Center section. With permission of Dietmar Feichtinger Architectes.

57

entrance canopy at one end of the building is composed of a large sloped cantilever. Unlike many other cantilevered canopies which function as only canopies, the cantilevered canopy in the Voest Alpine Office Center integrates interior programed space within it.

The structure of the building is composed of steel frames with concrete floors. Transversally, the space is organized as three zones, the central zone and two perimeter zones. The columns are located between the zones, and, consequently, the perimeter zones are all cantilevered. The proportion of the spans of one perimeter zone, central zone and the other perimeter zone is 1:3:1, which is close to the optimal. This proportioning makes the floor structure perform very efficiently. (See Figure 3-4 in Chapter 3.) The two cantilevered perimeter zones are occupied by individual offices and the central zone contains the vertical circulation cores, atrium and other shared functions, such as meeting areas, copy rooms, etc.

towards the entrance, the regularly placed columns stop, and the large cantilever begins from the third floor level and slopes up to terminate with the sharp edge. In conjunction with the façades set back on the ground floor facing the plaza, the hovering expression of the curvilinear mass with the sloping pointed cantilever becomes more dramatic. Two cantilevered trusses

Figure 2-36. Construction of Voest Alpine Office Center. Dietmar Feichtinger Architectes, Photo: Josef Pausch.

along the column lines of the building support the large cantilever. The trusses begin much earlier than the beginning of the cantilever to provide strength and stiffness necessary to support the cantilever. The tapered form produced by sloping up the cantilever corresponds to the structural logic of the cantilever. The expression of tapering the cantilever also begins much earlier than the actual beginning of the cantilever. Combined with the glass façade design of the entrance lobby under the cantilever and set-back façades, the length of the cantilever looks longer than actual. Trusses are placed not only vertically to carry the gravity loads of the cantilever but also laterally to carry the lateral loads applied to the cantilever as can be seen in the construction photo of the building.

CHAPTER 3
CANTILEVERED BUILDINGS

WHILE SMALLER SCALE CANTILEVERS have been used in buildings throughout the history of architecture, large scale dramatic cantilevers are relatively new. Traditional building materials, such as stone and wood, have critical limitations to produce large cantilevers. Structural materials for large cantilevers must be strong in tension and compression for safety because the load carrying mechanism of cantilevered structures requires both tensile and compressive strength. In addition, they should be stiff enough to prevent excessive deformations for serviceability. While certain types of stone have significant compressive strength, no stone has substantially reliable tensile strength. Wood has both compressive and tensile capacity if used in a proper direction. However, for very large cantilevers, the strength of wood may not be sufficient. In addition, wood may not be stiff enough to control displacements of very large cantilevers.

The emergence of large cantilevers was based on the use of iron and steel initially in the mid-19th century and reinforced concrete soon after. Large horizontal cantilevers were used first not for buildings but for bridges. The Hassfurt Bridge of 1867 in Germany with a central span of 38 m is generally recognized as the first modern cantilever bridge. Since then, many cantilever bridges have been built in steel or reinforced concrete throughout the world. One of the most well-known cantilever bridges is the Forth Bridge of 1890 in Scotland. The longest span of the bridge is 520 m, which is composed of two 207 m cantilevers and a 106 m central structure between them supported by the two tips of the cantilevers. The cantilever bridge having the longest single span at this time is the Quebec Bridge of 1919. The 549 m span is composed of two 177 m cantilevers and a 195 m central structure between them. Both bridges use steel trusses for the cantilevers and the central structures.

The lengths of the cantilevers in these long span cantilever bridges are significant. When considered in terms of building heights, the length of

Figure 3-1. Forth Bridge and its construction.

the cantilever of the Forth Bridge is equivalent to the height of a building of about 60 stories. However, in cantilever bridges, large cantilevers appear only during the construction process. In fact, one of the most important motivations for the development of cantilever bridges is that using cantilevers a long span bridge can be constructed without falsework. Cantilevers are used in bridges not as a final built form, but as a very efficient construction methodology. In order to satisfy the most important functional requirement of bridges, the cantilevers must be connected eventually. Therefore, once completed, large cantilevers no longer exist in cantilever bridges.

When large cantilevers are designed for buildings, they almost always remain as cantilevers. Large iron/steel cantilevers emerged in architecture in the late 19th century not as horizontal cantilevers but as vertical cantilevers, including the early tall buildings in Chicago and New York. Though not a typical building, the Eiffel Tower of 1889 in Paris is one of the most renowned vertical cantilevers even today. The tower is 300 m tall and constructed with wrought iron. Against wind loads, the tower has a structurally logical form following

Figure 3-2. Reinforced concrete cantilever bridges under construction. With permission of The Soletanche Freyssinet Group.

the bending moment diagram of a vertical cantilever beam subjected to lateral loads. Design and performance of vertical cantilevers will be discussed in greater detail in Part II of this book. If used for horizontal cantilevers with a 90-degree turn, the tower would be transformed to a large one-sided cantilever and the tapered form would still be logical now to carry gravity loads. However, for horizontal cantilevers in buildings, a non-tapered form is preferred in general for better functional performance. Furthermore, a horizontal cantilever of the size of the Eiffel Tower would require a gigantic

Figure 3-3. Eiffel Tower (left) and Eiffel Tower imaginarily used as a horizontal cantilever (right).

vertical structure and foundation to support it. Designing and building a large horizontal cantilever including its vertical support and foundation system are very challenging tasks.

This chapter presents the concept of horizontal cantilevers employed for a significant portion of a building. It begins with discussions on efficient proportioning of symmetrical and asymmetrical cantilevers. After that, cantilevered buildings of various configurations are presented, such as one-sided cantilevers, two-sided cantilevers, merged cantilevers and stacked multiple cantilevers. For each category, structural concepts are introduced first and their applications to real world examples are presented. In many cases, alternative design scenarios are comparatively studied to simulate typical design processes and understand how buildings with large cantilevers of alternative configurations perform.

3.1. CANTILEVER PROPORTIONING

Lightness is one of the most important themes of modern architecture and architects have pursued various design ideas to achieve it. One of the most effective strategies to obtain physical lightness is lightening building structures by employing appropriately proportioned cantilevers. When a beam type structure with two simple supports, subjected to uniformly distributed loads, is considered, the optimal locations of the supports to produce the lightest structure are not at the two ends of the structure. This configuration makes the entire structure bend downwards, and consequently positive bending moments are developed throughout the structure with the maximum at the mid-span (see the first bending moment diagram of Figure 3-4). When this beam type structure is designed with a prismatic member, the maximum bending moment governs the design. This common design approach results in structural inefficiency because the beam is overdesigned except for the maximum bending moment portion at the mid-span.

As the two simple supports at the ends begin to be pushed in by the same distance, two symmetrical cantilevers are produced outside the supports. As the free ends of the cantilevers and the central span bend downwards, the two support regions bend comparatively upwards. Consequently, negative bending moments are developed around the supports with the negative maximum at the supports, and positive bending moments are developed between the inflection points in the central span with the positive maximum at the mid-span. The absolute values of these maximum negative and positive bending moments are smaller than that of the maximum positive bending moment of the original simply supported beam type structure with two end supports. As the two supports are continuously pushed in, the maximum negative bending moment becomes larger and the maximum positive bending moment becomes smaller (see the second through fourth bending moment diagrams of Figure 3-4). When the length of the two end cantilevers reaches about 21 percent of the entire length of the beam type structure,

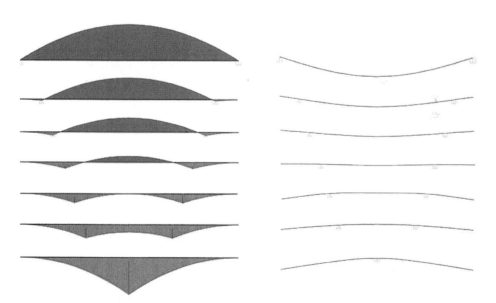

Figure 3-4. Comparative bending moment diagrams and deformed shapes of a beam type structure subjected to uniformly distributed loads and symmetrically supported by two simple supports of various locations.

65

the absolute values of the increasing maximum negative and decreasing maximum positive bending moments become the same (see the fourth bending moment diagram of Figure 3-4). This is the structurally optimal condition, which makes it possible to build the beam type structure with the lightest structural member.

As the two end supports are further pushed in after passing through the optimal locations, the maximum negative bending moments at the supports are continuously increased, and the maximum positive bending moment at the mid-span is continuously decreased and reaches zero when the length of the two end cantilevers is 25 percent of the entire length of the structure (see the fifth bending moment diagram of Figure 3-4). Once the support locations pass through the 25 percent points, the entire structure is subjected to negative bending moments (see the sixth bending moment diagram of Figure 3-4). When the two supports eventually merge at the mid-span, the maximum negative bending moment is developed there. The absolute value of this maximum negative bending moment is the same as that of the maximum positive bending moment of the original simply supported beam type structure with two end supports (see the last bending moment diagram of Figure 3-4).

Figure 3-4 also shows comparative deformed shapes of the beam type structures discussed thus far. As the end supports are continuously pushed in to the optimal locations, the deformation of the structure, which represents its serviceability, is gradually reduced (see the first through fourth deformed shapes). After passing through the optimal locations, the deformation becomes larger than the optimal condition. When the two supports are eventually merged, the displacements of the free ends of the cantilevers become their maximum (see the fifth through last deformed shapes). This maximum displacement is still smaller than that of the first case with two end supports, by about 40 percent.

Figure 3-5. New National Gallery in Berlin. With permission of Manuela Martin.

Figure 3-6. Cantilever barn in East Tennessee.

In conclusion, when appropriate, cantilevers can be employed to make structures lighter. For example, in the New National Gallery in Berlin designed by Ludwig Mies van der Rohe shown in Figure 3-5, each perimeter beam of the roof structure is symmetrically supported by two exterior columns which divide the beam into two cantilevered zones and the central span zone between the columns. This configuration produces the desired architectural form and, at the same time, a more efficient structural solution by reducing the maximum bending moment compared with the case with two end columns with no cantilevers. Another example is shown in Figure 3-6. The cantilevered upper portion of the cantilever barns in East Tennessee is also proportioned to reduce the maximum bending moment.

When only one support is pushed in and the other support remains at the end, only one cantilever is created beyond the pushed-in support. This asymmetrical support condition can also result in more efficient structural performance with reduced bending moments compared with the original simply supported beam type structure with two end supports. As the free end of the cantilever and the back span between the supports bend downwards, the cantilever side support region bends comparatively upwards. Consequently, negative bending moments are developed around the cantilever side support with the negative maximum at the support, and positive bending moments are developed in the back span between the end support and the inflection point with the positive maximum at the center of the positive

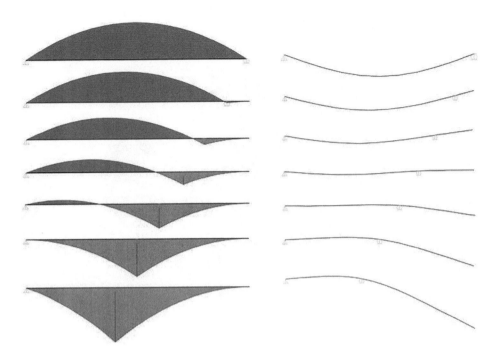

Figure 3-7. Comparative bending moment diagrams and deformed shapes of a beam type structure subjected to uniformly distributed loads and asymmetrically supported by one end support and the other support of various locations.

moment region. The absolute values of these maximum negative and positive bending moments are smaller than that of the maximum positive bending moment of the original simply supported beam type structure. As the cantilever side support is continuously pushed in, the maximum negative bending moment becomes larger and the maximum positive bending moment becomes smaller (see the second through fourth bending moment diagrams of Figure 3-7). When the length of the cantilever reaches about 29 percent of the entire length of the beam type structure, the absolute values of the increasing maximum negative and decreasing maximum positive bending moments become the same (see the fourth bending moment diagram of Figure 3-7). This is the structurally optimal condition, which makes it possible to build the beam type structure with the lightest structural member, for this asymmetrical configuration.

As the cantilever side support is further pushed in after passing through the optimal location, the maximum negative bending moment at the support is continuously increased and the maximum positive bending moment in the back span is continuously decreased and reaches zero when the length of the cantilever is 50 percent of the entire length of the structure (see the fifth and sixth bending moment diagrams of Figure 3-7). The absolute value of the maximum negative bending moment in this condition is the same as that of the maximum positive bending moment of the original simply supported beam type structure. Once the cantilever side support passes through the 50 percent point, the entire structure is subjected to negative bending moments, with the absolute maximum value larger than the maximum positive bending moment of the original simply supported beam type structure. Therefore, as the length of the cantilever becomes larger than 50 percent of the entire length of the beam type structure, the system is less efficient than the original case. When the two supports eventually meet at the end of the structure, the merged support must be changed to a rigid support for the resulting structure to be stable. In this case, a pure cantilever with no back span is obtained and the absolute value of the maximum negative bending moment at the support is four times larger than that of the maximum positive bending moment in the original simply supported beam type structure.

Figure 3-7 also shows comparative deformed shapes of the beam type structures discussed thus far. As one end support is continuously pushed in to the optimal location, the deformation of the structure, which represents its serviceability, is gradually reduced (see the first through fourth deformed shapes of Figure 3-7). After passing through the optimal location, the deformation becomes larger than the optimal condition. Once the pushed-in cantilever side support reaches the 50 percent point of the beam type structure, the entire structure bows upwards and is subjected to only negative bending moments. The maximum displacement of the free end of the cantilever in this case is larger by about 20 percent than that of the mid-span of the first case with two end supports. When the two supports are eventually merged at the end of the structure as a rigid support, the maximum displacement of the free end of the cantilever is larger by about 10 times than that of the mid-span of the first case without cantilever.

When appropriate, asymmetrical cantilevers can be used to make structures lighter. Figure 3-8 shows a naturally created asymmetrical cantilever structure, which is very similar to what have just been discussed. The Carne House in the Rheinauhafen of Cologne designed by Hadi Teherani of BRT Architecten, discussed in detail later in this chapter, is also a good example of asymmetrically cantilevered buildings with efficient structural proportions.

In the symmetrically supported beam type structure with two supports and subjected to uniformly distributed loads shown in Figure 3-4, both support reactions are acting upward and equal, regardless of the proportions of the cantilevers. Therefore, the typical vertical support system composed of columns and foundations is subjected to compression. In the asymmetrically supported beam type structure shown in Figure 3-7, as the right support is pushed in to create a cantilever, the left end support's upward reaction becomes reduced, while the cantilever side right support's upward reaction becomes increased (see the first and second diagrams

Figure 3-8. Asymmetrical cantilever structure in nature. With permission of Imp Adventures – Damon Blackband.

of Figure 3-9). When the cantilever side support is positioned exactly at the center of the beam type structure, the magnitude of the center support's upward reaction becomes the same as the sum of the downward uniformly distributed applied loads, and the left end support does not carry any vertical load (see the third diagram of Figure 3-9).

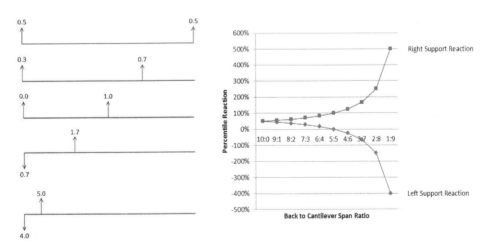

Figure 3-9. Relative support reactions of a beam type structure subjected to uniformly distributed loads and asymmetrically supported by one end support and the other support of various locations.

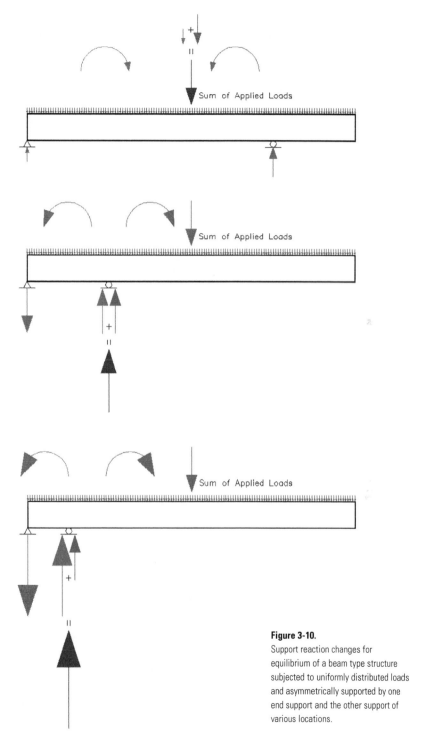

Figure 3-10.
Support reaction changes for equilibrium of a beam type structure subjected to uniformly distributed loads and asymmetrically supported by one end support and the other support of various locations.

71

As the cantilever side right support passes through the center of the structure and gets closer to the left end support, a downward reactional force begins to be developed in the left end support to prevent overturning failure and the cantilever side support's upward force becomes even greater than the sum of the downward forces applied to the beam type structure. This tendency becomes greater as the cantilever side support gets closer to the left end support (see the fourth and fifth diagrams of Figure 3-9 and second and third diagrams of Figure 3-10). Apparently, the expression of the cantilever becomes more dramatic as the back span length becomes shorter. However, the cantilever, back span and vertical supports are subjected to greater stresses, and consequently a more expensive solution is required for the structure. Furthermore, the development of a downward reactional force at the left end support means that the vertical support system is subjected to tensile forces. Development of tensile forces in the foundation system is not structurally desirable because soil does not have tensile resistance. A deep foundation system which carries tensile forces by frictions or an unusually large foundation system with an appropriate configuration to resist this tensile force is required to resolve this structural issue. Examples of this are the very large foundation systems of the Trumpf Campus Gate House presented in Chapter 2 and the former Lamar Construction Company Corporate Headquarters to be discussed in this chapter. Dramatically proportioned long cantilevers with short back spans typically require very expensive superstructures as well as substructures. Therefore, the proportions of cantilevered structures should be carefully configured considering not only expressed portions but also their foundation systems and related cost issues.

3.2. JETTYING

Jettying in wood structures has been a very practical and efficient method to increase the occupiable area of buildings by cantilevering floors. A jetty is produced by projecting upper floor beams or joists beyond the load-bearing walls of the floor below. In three-story wood frame structures, jettying can be used twice so the third floor is larger than the second floor and the second floor is larger than the ground floor. Jettying can be used for not only one side but also multiple sides of the building. When jettying is used for two sides meeting at a right angle, a diagonal dragon beam may be used to project the floor joists around the corner, beginning from the dragon beam.

In addition to increasing the occupiable floor area, jettying can produce a more efficient structural solution. Compared with the case with simply supported joists between the load-bearing walls, cantilevered joists having an even longer total length can be subjected to smaller bending moments. Let's consider a symmetrically cantilevered joist beyond the load-bearing walls. When the joist is subjected to only uniformly distributed gravity loads, the

Figure 3-11. Jettying in a wood structure.

maximum bending moment of the cantilevered joist can be smaller than that of the simply supported non-cantilevered joist. When the cantilevered length of the joist to the both sides is about 36 percent of the main span between the supports, the proportion of which is close to optimal, the maximum bending moment is only about 50 percent of the case with no cantilever.

When jettying is used to cantilever the second floor of a two-story building, however, the proportion of the cantilevered length should be much smaller in order to achieve the same structural efficiency because the joist is not only subjected to distributed loads from the floor but also large point loads at its two ends from the second floor exterior walls and roof. (No interior loadbearing walls are considered here for clearer conceptual discussions.) Therefore, if the 36 percent cantilevered length is still used and typical light

73

Figure 3-12.
Jettying at a building corner
with a diagonal beam.

wood frame walls, roof and floor loads are considered, the maximum bending moment of the joist can be, in fact, increased to about 200 percent of the case with no cantilevers as can be seen in Figure 3-14. The entire joist bows upwards and all negative bending moments are developed along the joist in this case because the two large point loads push down the ends of the cantilevers. When no jettying is used, the second-floor exterior wall and roof loads are directly carried by the vertical load bearing walls, and the second-floor joists carry only floor loads.

In order to reduce the maximum bending moment of the second-floor joists by about 50 percent in the symmetrically jettied two story building, the length of the cantilevers should be limited to about 8–10 percent of the main span length, when typical light wood frame loads are considered. With these shorter cantilevers in combination with the large point loads at their tips and uniformly distributed loads over the joist, the positive bending moment at the mid-span and negative bending moment at the supports are balanced, and structural efficiency is obtained again.

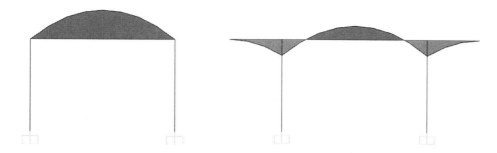

Figure 3-13. Comparative bending moment diagrams of simply supported and optimally cantilevered joists with the same primary span length between the two vertical supports.

When the symmetrically cantilevered length of the second-floor joists is about 21–23 percent of the main span length, the maximum bending moment of the joists is about the same as that in the simply supported case with no cantilevers. While the simply supported case develops all positive bending moments throughout the joist, this case with 21–23 percent cantilevers develops almost all negative bending moments because of the pushing down effects caused by the large point loads at the tips of the cantilevers. When the length of the cantilevers is greater than about 23 percent, a larger maximum bending moment is developed than that in the case with no cantilevers.

Jettying can be used to cantilever only one end of the joist. When only uniformly distributed loads are applied to the joist, the structurally optimal

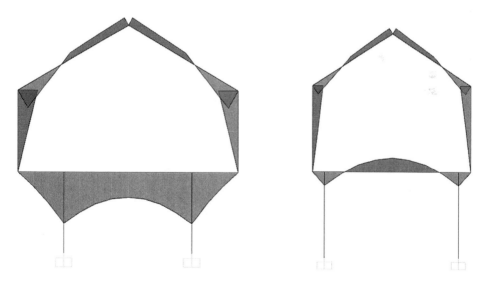

Figure 3-14. Comparative bending moment diagrams of symmetrically jettied two-story buildings with the same main span length between the two vertical supports and different cantilevered lengths.

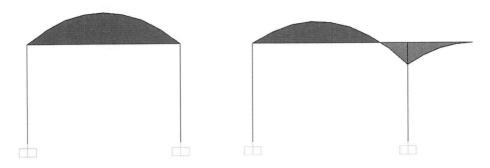

Figure 3-15. Comparative bending moment diagrams of simply supported and optimally one-sided cantilevered joists with the same primary span length between the two vertical supports.

length of the cantilever is about 41 percent of the main span between the load-bearing walls. Based on this proportion, the maximum bending moment of the joist is reduced to about 70 percent of that in the case with no cantilever. This reduction is caused by balancing the all positive bending moments of the simply supported case into positive between the end support and inflection point and negative around the cantilever side support as can be seen in Figure 3-15.

When this one-sided cantilever is used for a two-story building to increase the occupiable space of the second floor, the 41 percent cantilever with typical light wood frame loads can result in a large increased maximum bending moment of about 300 percent of the case with no cantilever. This is because the joist is subject to not only distributed loads from the floor but also a large point load applied at the tip of the cantilever from the second-floor exterior wall and the roof. Due to the large point load, almost the entire joist develops negative bending moments.

In order to reduce the maximum bending moment of the second-floor joists of the two-story building with a one-sided cantilever, the proportion of the cantilever should be reduced. With a cantilevered length of about 18 percent of the main span, the system is structurally optimized. In this case, the maximum bending moment is reduced to about 70 percent of the case with no cantilever.

When the one-sided cantilever length of the second-floor joist is about 25 percent of the main span length, the maximum bending moment is about the same as that in the case with no cantilever. Therefore, the occupiable area of the second floor can be increased by 25 percent without increasing the maximum bending moment of the joists. The same size joists used for the case with no cantilever can still be used for the jettied building with the second-floor area increased by 25 percent.

The Shambles in York, England is well known for jettied buildings. Figure 3-17 shows three-story jettied buildings towards the street. As can be

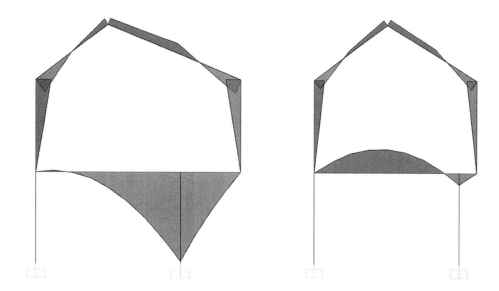

Figure 3-16. Comparative bending moment diagrams of asymmetrically jettied two-story building with the same primary span length between the two vertical supports and different cantilevered lengths

Figure 3-17.
The Shambles in York, England.

77

seen in the figure, the third floor is larger than the second floor, and the second floor is larger than the ground floor. By this technique, the width of the street is maintained, while the floor area of the upper levels of the buildings along the street is maximized.

In order to achieve the optimal structural efficiency in three-story buildings with jettied second and third floors, the proportion of the cantilevered lengths of the second and third floor should be carefully determined. When symmetrical jettying is used as can be seen in Figure 3-18, the cantilevered length of the second-floor joist should be limited to about 4 percent of the first-floor span length because the point loads applied to the tips of the second-floor cantilevered joists by the second and third floor exterior walls, third floor, and roof are very large. The cantilevered length of the third-floor joist can be increased to 8–10 percent of the second-floor span length because the point loads applied to the tips of the third-floor joists by only the third-floor exterior walls and roof are much smaller than those applied to the tips of the second floor joists. By these proportions, the second and third floor joists are structurally optimized to develop the minimized maximum bending moments. In this example, the main span of the third floor between the vertical supports is larger than that of the second floor. Therefore, the optimized maximum

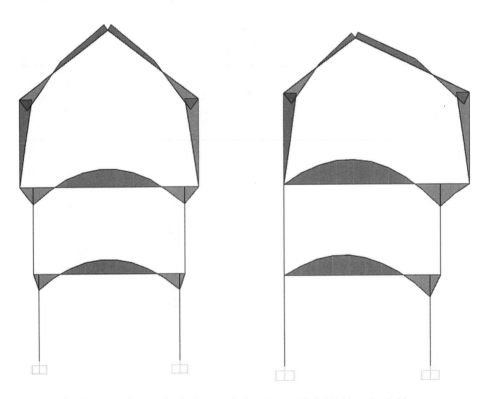

Figure 3-18. Bending moment diagrams of optimally symmetrically and asymmetrically jetted three-story buildings

bending moment of the third-floor joists is larger than that of the second-floor joists.

When one-sided jettying is used, the cantilevered length of the second-floor joist should be limited to about 7 percent of the first-floor span length because the point loads applied to the tips of the second-floor cantilevered joists by the second and third floor exterior walls, third floor, and roof are very large. The cantilevered length of the third-floor joists can be increased to about 18 percent of the second-floor span length because the point loads, applied to the tips of the third-floor joists by only the third-floor exterior walls and roof, are much smaller than those applied to the tips of the second floor joists.

Between the symmetrically and asymmetrically jettied buildings, the symmetrical one creates additional floor area more efficiently. The same amount of floor area can be added by symmetrical jettying with less amount of joist material because smaller bending moments are developed in the symmetrical configuration.

3.3. LARGE ONE-SIDED CANTILEVERS

One of the most challenging configurations of cantilever is a large one-sided cantilever, especially with a short back span. When this type of configuration is employed as an option to solve certain design problems, its architectural

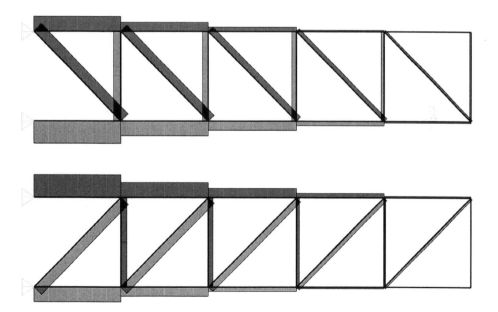

Figure 3-19. Axial forces of cantilevered trusses (tension in darker shade and compression in lighter shade in all axial force diagrams in this chapter).

expression can be very dramatic, while its structural solution is demanding. The structural performance of one-sided cantilevers is dependent on various factors, such as the type and depth of the structural system, proportion between the cantilevered and back spans, and configuration of the foundation system.

Trusses are one of the most predominantly used structural systems for large cantilevers in buildings. By continuously connecting linear structural members in triangular forms, trusses carry applied loads at the nodes very efficiently by the component members' axial actions. Therefore, while the system is typically very light, it is very strong and stiff. Figure 3-19 shows cantilevered trusses of two different configurations. In cantilevered trusses, the top and bottom chord members develop tensile and compressive forces respectively to resist the gravity-induced overall bending of the system. The axial forces of the chord members become larger towards the support, where the maximum bending moment of the cantilever is developed.

The web members between the top and bottom chord members are typically composed of either vertical or slanted members to form triangular shapes in the truss and primarily carry the overall shear force of the cantilever. Depending on the configuration, the shear of the cantilever is carried by either tension or compression of the web members. It is often preferred to configure

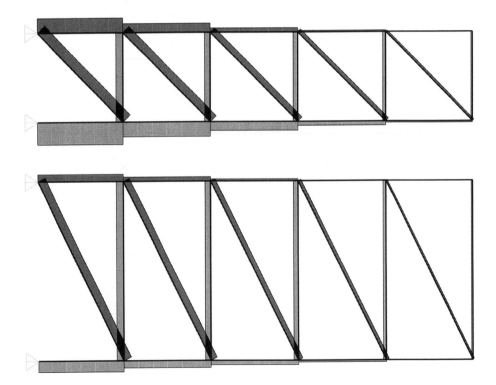

Figure 3-20. Comparative axial forces of cantilevered trusses of different depths.

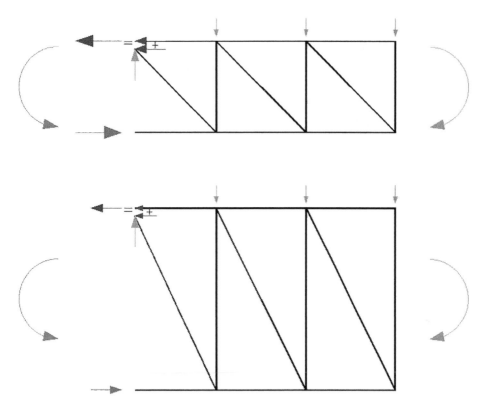

Figure 3-21. External and internal force conditions of the trusses of different depths.

the web members in such a way that longer diagonal members develop tensile forces and shorter vertical members develop compressive forces as shown in the upper diagram of Figure 3-19. This is because long and slender compressive members are vulnerable to buckling failure. Therefore, strategically making longer members subjected to tensile forces can eliminate the possibility of their premature buckling failure. The axial forces of the web members also increase towards the support because the overall shear force of the cantilever becomes larger towards the support. The slanted web members also participate in carrying the overall bending of the cantilever in association with the previously discussed top and bottom chord members which primarily resist the overall bending of the system.

Figure 3-20 shows two cantilevered trusses of different depths subjected to the same gravity loads on the nodes. As the depth of the truss becomes larger, the top and bottom chord member forces are reduced. When the depth of the truss becomes twice to carry the same gravity load, the top and bottom chord member forces become about half. This is because the top and bottom chord members in combination with the horizontal

81

components of the diagonal members in the web produce resisting moment by their axial actions against the overall bending of the cantilever as can be seen in Figure 3-21. The depth of the truss works as the resisting moment arm. As the length of the resisting moment arm becomes twice, the top and bottom chord member forces become half to carry the same overall bending moment. Therefore, as the length of the cantilever becomes large, it is a good strategy to develop a story height or even two or more story height truss system to support the cantilever more efficiently with a maximized resisting moment arm. When these deep trusses are employed on the façade planes of cantilevered buildings, visual and environmental connections between the interior and exterior can still be obtained through the triangular openings. When they are used in the interior space, circulations can also be obtained again through the triangular openings.

Considering the fact that the overall bending moment of the canti-levered system is increased towards the support, a structurally corresponding more logical form of the cantilevered trusses can be conceived. When external gravity loads are applied to the nodes of the cantilevered truss, the cor-responding overall bending moment of the system is shown in the upper diagram of Figure 3-22. If the shape of a cantilevered truss follows the form of this bending moment diagram, horizontal components of every top and bottom chord member forces which carry the overall bending moment of the system become identical. This is because in the reshaped truss the resist-ing moment arm changes in proportion to the required magnitude of resisting moment along the cantilever. Therefore, the horizontal members of the truss, which develop only horizontal forces, can be designed with all identical members. The horizontal components of the slanted member forces are also all identical. The varying depths between the top and bottom chord members, which represent the varying resisting moment arms, in combination with the constant horizontal top and bottom chord member forces, reflect the required varying moment resistance of the system at different locations. The vertical components of the slanted member forces are what carry the shear forces of the system. As the shear force increases towards the support, the slanted member size should be increased accordingly towards the supports. The lower diagram of Figure 3-22 shows axial forces of a so-called funicular-shaped canti-levered truss following the structural logic described above. Once shaped in this way, no internal forces are developed in the web members as can be observed in the figure.

Deep trusses of triangular geometric configurations produce very efficient structural solutions for large cantilevers. However, this strategy may involve large diagonal members one or more stories tall. Inclusion of diagonal members and consequent triangulation can produce superior structural solu-tions in terms of both strength and stiffness. However, large diagonal members in buildings are not always welcomed architecturally. If diagonal members must be prevented, Vierendeel trusses can be considered as an alternative design

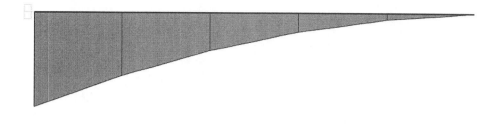

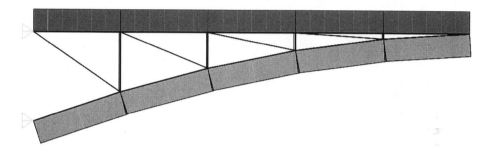

Figure 3-22. Overall bending moment and axial force diagrams of funicular-shaped cantilevered truss.

solution. Once all the diagonal members are eliminated from the conventional parallel chord cantilever truss with vertical and diagonal web members and all the connections are changed to moment connections, a Vierendeel cantilever is produced.

With its orthogonal configuration, the cantilevered Vierendeel truss carries the applied loads no longer by only axial actions. The overall bending of the cantilever is still carried by axial actions, tension in the top chord members and compression in the bottom chord members. The axial forces become larger towards the support of the cantilever because overall bending of the cantilevered structure increases towards the support. However, these axial forces typically do not govern the structural design of the Vierendeel system. With no diagonal members, shear forces of the system are carried by bending of the top and bottom chord members as well as vertical members. Since shear forces of the cantilever increase towards the support, the bending moments of the top and bottom chord members as well as vertical members to carry the shear forces also increase towards the support.

When axial and bending actions work together in Vierendeel truss members, bending typically governs the structural performance and design of the system. Bending action is a very inefficient load carrying mechanism.

83

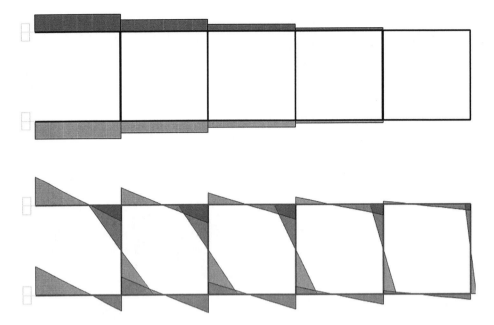

Figure 3-23. Axial forces and bending moments of cantilevered Vierendeel truss members.

(It is not too difficult to recognize that axial action is a much more efficient way of carrying forces than bending. For example, a bamboo chopstick can be easily broken by bending even by the hands of a child, but it is almost impossible to break it by pulling or pushing it from both ends axially even for a strong adult person, which explains the relative efficiency of axially loaded structural members.) Compared to the conventional triangulated truss system, the Vierendeel system, the design of which is primarily governed by each member's bending action, requires much larger member sizes to provide sufficient strength and stiffness. Figure 3-24 shows comparative deformed shapes of the conventional triangulated cantilevered truss and Vierendeel cantilever designed with the same amount of structural material and subjected to the identical gravity loads at the nodes. Significantly larger deformations are observed in the Vierendeel cantilever. Therefore, Vierendeel trusses should be considered carefully perhaps only for cantilevers of relatively short length when they are very much required.

Performance of a building with a large one-sided cantilever is also greatly affected by the proportion between the back span and cantilever. When there is only one back span, the optimal back span to cantilever length ratio of the structure subjected to uniformly distributed loads is about 10:4 as was discussed earlier. When there are multiple back spans of equal length, the optimal proportion should primarily be considered only with the immediate back span of the cantilever, not with the combined length of the back spans.

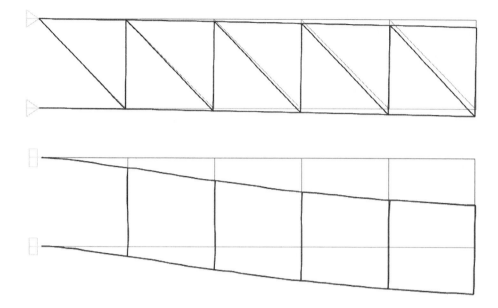

Figure 3-24. Comparative deformed shapes of cantilevered truss and Vierendeel cantilever.

The first bending moment diagram of Figure 3-25 shows the case of the optimally proportioned one-sided cantilever with a single back span. As the number of the back span supports is increased to divide the back span into equal length multiple back spans, the bending moments of the back spans, excluding the first back span immediately after the cantilever, are decreased. However, the bending moments of the cantilever and the first back span do not change much.

In Figure 3-25, as the number of back span supports is increased, the proportion of the cantilever to the first back span length immediately after the cantilever is increased. In the fourth moment diagram with four back spans, the first back span length is smaller than the cantilever length. In this case, the second support from the cantilever develops a downward reaction force. Therefore, tensile force is expected in the column and the foundation system which supports this portion of the structure. Development of tensile forces in the foundation system is not desirable structurally because soil does not provide tensile resistance. In the fifth diagram, even though the first back span length immediately after the cantilever is still shorter than the cantilever length, the downward reaction force of the second support from the cantilever is eliminated by increasing the length of the second back span. Figure 3-26 shows percentile support reaction forces of the structures shown in Figure 3-25.

The sixth diagram of Figure 3-25 shows the case with the optimally reduced cantilever length based on the first back span length when there are

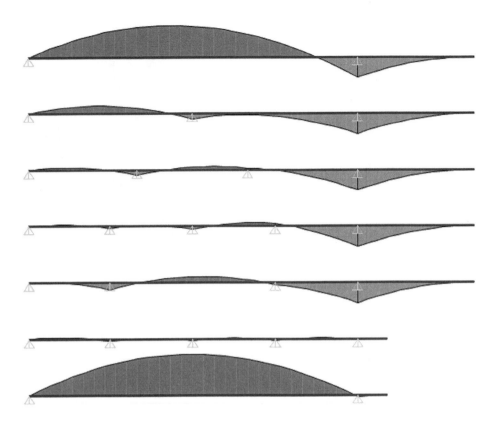

Figure 3-25. Bending moment diagrams of one-sided cantilevers with single and multiple back spans of various configurations.

multiple equal length back spans. Bending moments of every span are mini-mized and downward reaction forces are no longer developed in any support. If the four back spans are combined as a longer single back span as can be seen in the last diagram, bending moment of the combined back span between the supports becomes even larger than the first case with a longer but optimized length cantilever. In structures with a large one-sided cantilever, not only the absolute length of the cantilever but also the proportional relationship between the back spans and the cantilever is a very important factor for efficient structural performance.

The primary load to be considered in large one-sided cantilevers is typically gravity as has been discussed thus far. When the cantilever is very long and slender, however, lateral loads also significantly influence its structural design and performance. While gravity loads always have the pre-determined direction, lateral loads should be considered in any direction because wind can blow in any direction. In steel structures, diagonal cross bracings can be employed on the floors of the cantilever as an effective means

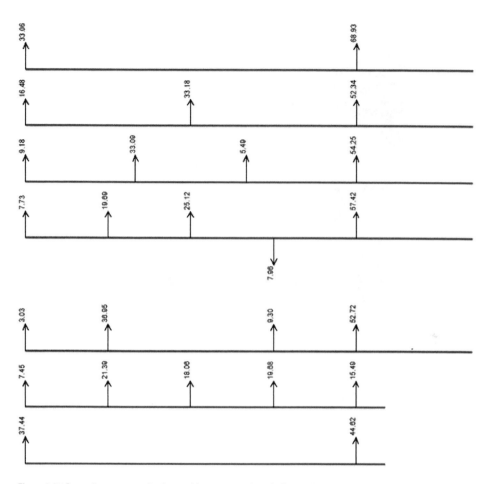

Figure 3-26. Percentile support reaction forces of the structures shown in Figure 3-25.

to carry lateral loads. In reinforced concrete structures, the floor structure should be designed not only as the gravity load carrying system but also as the lateral load resisting diaphragm.

ICA Boston, USA

The Institute of Contemporary Art (ICA) building in Boston designed by Diller Scopidio + Renfro is dramatically integrated with Boston Harbor with a very large one-sided cantilever. The three floors from the ground have a smaller footprint due to the limited size of the site by the waterfront, while the fourth-floor gallery space of a much larger footprint is created by the long one-sided cantilever of about 82 ft (25 m) hovering over the Boston HarborWalk next to the water. The resulting profile of the building based on this configuration is an inverted L with an impressive cantilever.

Figure 3-27. The Institute of Contemporary Art (ICA) in Boston. Photo by Iwan Baan, Courtesy of DS+R.

Four trusses of the full depth of the fourth floor, which is about 23 ft (7 m) , support the large cantilever. The trusses are continued from the entire back span to the cantilever to provide sufficient strength and stiffness. Each truss is composed of seven modules. With primarily two columns supporting each trusses – one at the end of the truss and the other four modules apart from the end column – the back span to cantilever length ratio is 4:3, which is not structurally optimal. However, this proportion creates a very dramatic cantilever, and, when identical loads are applied to the nodes of the truss, only compressive forces are developed in the columns and consequently in the foundation.

In fact, there are more columns between the two primary truss-supporting columns to carry the loads from the three floors on the east side of the building under the fourth floor gallery. However, sliding joins are used between these columns and the trusses so that the gravity loads from the trusses are not directly carried by these columns (see the first diagram of Figure 3-28). If these interior columns are directly connected to the trusses, some of the columns will be subject to tensile force development. On the west side of the building, the performance theater is hung from the perimeter truss by tubular steel hangers (see the second diagram of Figure 3-28). With this hanging configuration, the end column is less vulnerable to tensile force development even in cases with relatively large live loads only in the canti-levered portion. Furthermore, based on this hanging design, column-free lobby space on the ground floor is provided under the theater.

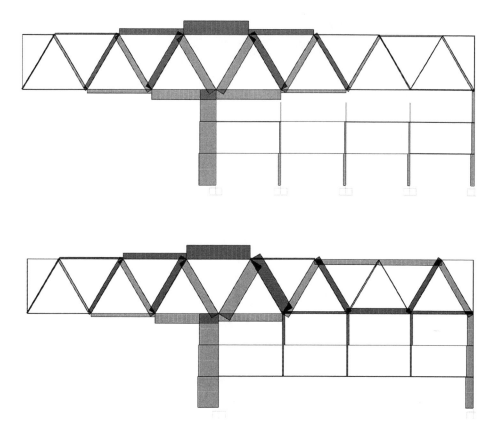

Figure 3-28. Comparative axial force diagrams of simplified ICA structure in Boston (top: east side frame, bottom: west side frame).

A proportion close to the optimal would be created by cantilevering only two modules of the trusses instead of three. With the back span to cantilever length ratio of 5:2, the overall member forces of the trusses are minimized, and consequently the trusses can be constructed with lighter members. However, the architectural expression of the cantilever of this proportion would be less dramatic. In addition, the HarborWalk outdoor space in front of the building would be reduced, while the interior space would be increased.

On the contrary, if the length of the cantilever were increased to four modules of the trusses, its hovering expression could be more significant. However, with the back span to cantilever length ratio of 3:4, the overall member forces of the trusses becomes much larger and the member sizes of the trusses should also be larger. In addition, with this proportion, tensile forces are developed in the end columns and consequently in the foundation system. Furthermore, compressive forces in the cantilever side columns become much larger compared with the 5:2 length ratio case. These tensile forces and very large compressive forces would result in very costly super- and

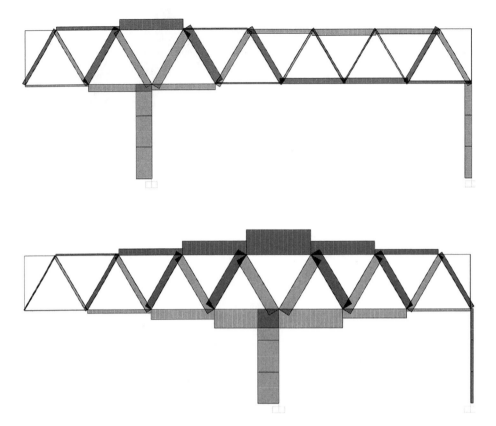

Figure 3-29. Axial force diagrams of the ICA building structure in Boston with alternative back span to cantilever length ratios.

sub-structures. As the proportion of the cantilever becomes larger, its deformation also becomes larger.

The actual foundation system for the ICA building is composed of steel H piles of longer than 100 ft (30 m), concrete pile caps and grids of concrete beams, which tie the pile caps. The piles carry not only compressive loads by end bearing and friction but also tensile loads by friction. The tensile capacity of the piles is limited to about 23 percent in this project. Based on the back span to cantilever length ratio of 4:3, it is not likely that tensile forces are developed in the end columns in typical cases. However, in a plausible extreme load case such as very large live loads only in the cantilevered portion, it is still possible for the end columns to develop some tensile forces, the magnitude of which will be much smaller than that of compressive forces developed in the other sets of columns at the beginning of the cantilever.

Two of the four mega-trusses are placed along the east and west side perimeter of the gallery and the other two are placed around the central core of the gallery. While the perimeter trusses are continued towards the free end of the cantilever on the north side, the trusses around the central core stop before they reach the north façade. In combination with mullion-less point fixing glass façades, this truss configuration helps create an uninterrupted spectacular water view to the north side of the building.

Depending on the façade design including material choices, the expression of cantilevers is significantly influenced. In the ICA Building, the same wood finish is continuously used from the walkway on the ground level under the cantilever up to the back of the theater with a 90-degree turn and returns onto the underside of the cantilever with another 90-degree turn. This band of same wood finish is clearly expressed on the building façade as an important design element. Most of the façade areas between the same wood finish bands are clad with transparent glasses, while the cantilevered gallery mass is clad with translucent material except for the north front façade facing water. This design strategy makes the expression of the cantilever with the back span to cantilever length ratio of 4:3 much more dramatic, as if the proportion of the cantilever were much greater than actual.

Former Lamar Construction Company Corporate Headquarters, Hudsonville, Michigan, USA

The former Lamar Construction Company Corporate Headquarters in Hudsonville, Michigan, designed by Integrated Architecture boasts one of the most dramatic cantilever structures in the US. In fact, the building does not necessarily require a large cantilever structure in terms of its site condition, program, etc. However, as the headquarters of a construction company, the building was designed to expressively present the company's ability to per-form steel erection and construction in a masterly way. The building is com-posed of two floors. Unlike typical two-story buildings, however, the second floor was not built on top of the first floor. Instead there is a very large gap between the first and second floor. In order to support the second floor which is hovering over the first floor, it is cantilevered to one side from the vertical reinforced concrete structural core containing a staircase and elevator.

The second floor cantilever is structured with two 112 ft (34.1 m) long steel trusses which are full story height of 16 ft (4.9 m) deep. At the junction, between the steel cantilever trusses and the vertical reinforced concrete core, where the bending moment of the cantilever is greatest, diagonal members are added under the trusses like brackets to increase the structural depth there. The diagonal members are geometrically integrated with the slope of the staircase between the first and hovering second floor. Comparative axial force diagrams of the cantilevered trusses with and without the additional diagonal members are shown in Figure 3-31. Reduced member forces can be clearly observed in the design with the increased structural depth.

With the very long cantilever and slender vertical core, the back span to cantilever length ratio is about 1:5 in this building. This proportion makes the building very vulnerable to overturning failure. To prevent the overturn-ing, the core structure is supported by a 90 ft (27.4 m) long, 62 ft (18.9 m)

Figure 3-30. Former Lamar Construction Company Corporate Headquarters. With permission of Van Dellen Steel, Inc.

wide and 6 ft (1.8 m) deep reinforced concrete footing, which is extended in the direction of the cantilever. The steel truss cantilever, reinforced concrete vertical core and unusually long, wide and deep underground footing create a C-shaped structure, which can effectively resist the overturning tendency of the building. The anticipated displacement of the large cantilever is

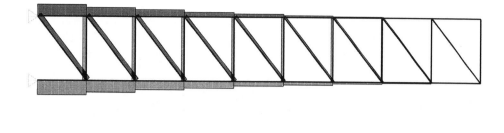

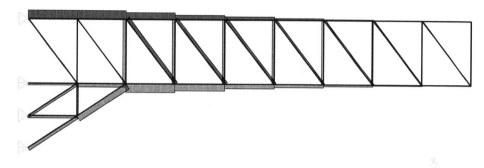

Figure 3-31. Former Lamar Construction Company Corporate Headquarters trusses of alternative configurations.

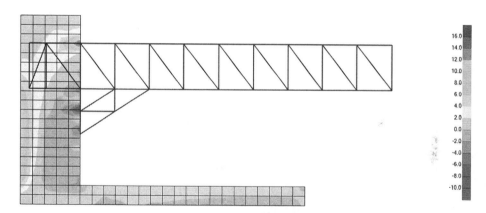

Figure 3-32. C-shape structural configuration of former Lamar Construction Company Corporate Headquarters.

significant. To compensate for that, a camber of 4 in (10.1 cm) was used at the free end of the cantilever, according to the project engineer.

The cantilevered portion of the building is composed of a rectangular box form volume which accommodates office space. The two large trusses which support the cantilever are located inside the cantilevered volume. The floor beams, supported by these two cantilevered trusses, are placed across the trusses and produce two symmetrical cantilevers. A plausible design

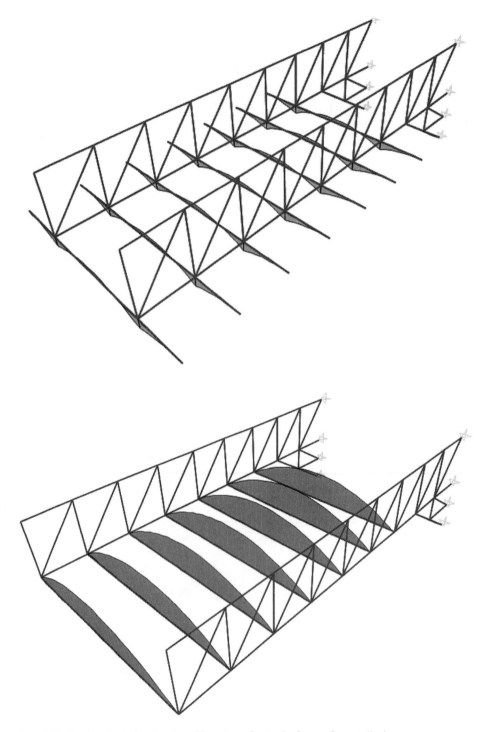

Figure 3-33. Alternative structural configurations of former Lamar Construction Company Corporate Headquarters.

alternative would be locating the two large trusses along the two longitudinal perimeter surfaces of the cantilevered rectangular volume and placing the floor beams between the trusses. In terms of structural performance, the actual construction is more efficient because the maximum bending moment of the floor beams with symmetrical cantilevers is smaller than that of the simply end-supported floor beams of the same total length. However, this configuration requires the trusses to be placed within the interior space and divides the space into three separate zones. Therefore, if the architecturally desired spatial configuration and this structural arrangement can be well integrated, this is a good design solution. In this building, the space between the trusses is used for the stairwell and conference room. The two symmetrically cantilevered zones beyond the trusses are used for offices, and the exposed trusses naturally define each space.

The scheme alternatively considered here can create a large column-free space. Therefore, if the architecturally required spatial organization prefers a large column-free space, this configuration can be a better design option, though the structural efficiency lacks compared with the scheme actually used. It should also be noted that the design alternative with perimeter cantilever trusses requires a design modification at the junction between the core and the cantilevered mass. A larger core is required to directly embed the cantilevered trusses, which will change the composition of the building masses and overall building aesthetics. Alternatively, double cantilevering can be used to keep the existing core size and maintain the current composition of the building masses.

The two different structural alternatives also affect the façade design. The constructed scheme does not require substantial structural elements on the building perimeter. Therefore, façade design can be performed with a great degree of flexibility, and, if desired, the transparency of the façade and visual connection between the interior and the exterior can be maximized. The alternative scheme requires placing large trusses on the building perimeter. Therefore, the façade design is significantly influenced by the structural elements. The heavy trusses may be visually exposed and, in turn, this may obstruct the view from inside. Considered from a different viewpoint, this situation could provide a good opportunity to express structures on the building façades as an important architectural design element when appropriate.

Milstein Hall at Cornell University, New York, USA

Milstein Hall designed by OMA is a new addition to the College of Art, Architecture and Planning of Cornell University. The new building, inserted into a very limited site between three existing buildings – Sibley Hall, Rand Hall and Foundry – makes the much desired direct connection between Rand Hall and Sibley Hall. The building's second floor, which mainly contains design studios, is lifted from the ground, hovers over the University Avenue by a large cantilever, and creates an indirect relationship with the Foundry.

Figure 3-34. Milstein Hall at Cornell University. Image courtesy OMA; Photography by Iwan Baan.

Four eccentrically configured trusses are employed for the 50 ft (15.2 m) long cantilevered portion over the University Avenue and Vierendeel trusses are used for the two back spans of about 90 ft (27.4 m) total. Both the cantilever and back span trusses are full story height deep. Between Sibley Hall and Rand Hall towards the back of the building, there is another narrower cantilever of about 50 ft (15.2 m) on the opposite side of the large main cantilever. This narrow cantilever is also structured with eccentrically configured full story height deep trusses.

Eccentrically configured trusses are not as strong and stiff as normal trusses of concentric triangular configurations. With greater ductility, they could perform better against seismic loads, but Cornell is not located in an active seismic zone. In Milstein Hall, the large full story height deep trusses are integrated with the studio space. The top and bottom chord members of the trusses are placed within the second and roof floor structures, while the slanted web members of the trusses are placed within the studio space. Eccentrically configured trusses with slanted web members not meeting each other at the chord members provide better circulations for the studio space, while they still provide an efficient structure for the large cantilever. For the back spans of the cantilever, the eccentrically configured trusses smoothly transform into Vierendeel trusses, which, without slanted members, provide even better circulations as well as functional and visual connections between the spaces. Certainly, the structural performance of the eccentrically configured trusses is superior to the Vierendeel trusses in terms of both strength and stiffness.

Figure 3-35. Cantilevered studio space of Milstein Hall at Cornell University. With permission of Philippe Ruault.

Figure 3-36 shows a simplified overall configuration of the structure and its bending moment and shear force diagrams. Negative bending moments in the cantilever are increased towards the support with the maximum at the support. Passing through the first supporting column, the maximum bending moment of the cantilever is shared by the column and the back span. Furthermore, inflection points exist in the back spans, and, consequently, absolute values of bending moments are much smaller in the back spans than those in the long cantilever. Shear forces are also large around the support of the large cantilever.

Considering the behavior of the simplified overall structure and functional issues including circulations, the actual structure was designed with eccentrically configured trusses for the long cantilevers, and with Vierendeel trusses for the back spans. Bending moments and shear forces are relatively small in the back spans. The maximum bending moment and shear force of the back spans are developed at the end of the span adjacent to the long cantilever. Considering this, the vertical members of the Vierendeel trusses there are slightly slanted instead of purely vertical.

For long cantilevers, it is better to employ normal concentric trusses. Compared with eccentrically configured trusses, concentric trusses produce stronger and stiffer structures. However, the eccentrically configured trusses in Milstein Hall provide more flexible interior space for the studio as has been discussed. If long cantilevers are designed with Vierendeel trusses instead of triangulated trusses to provide even more flexible interior space, it will

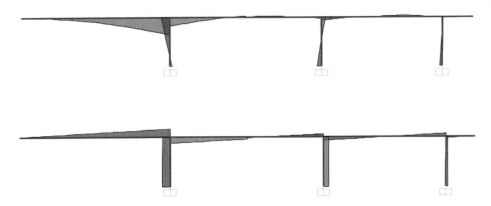

Figure 3-36. Bending moment and shear force diagrams of simplified Milstein Hall structural model.

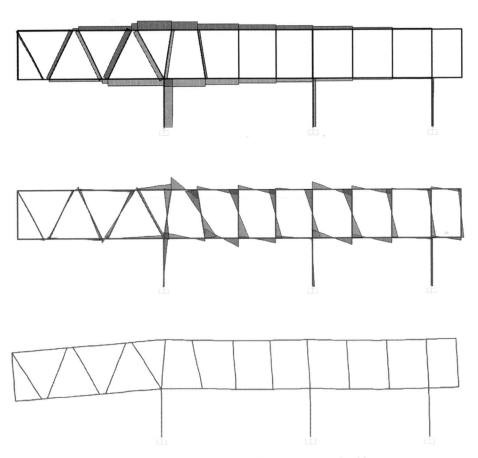

Figure 3-37. Axial force, bending moment and deformation diagrams of Milstein Hall structural model.

require much heavier structures to provide necessary strength and stiffness because Vierendeel trusses carry loads primarily by bending actions which are a very inefficient load-carrying mechanism.

Ataria Nature Interpretation Centre, Vitoria, Spain

The Ataria Nature Interpretation Centre in Vitoria, Spain, designed by estudio ATARIA is primarily built with wood, which, as one of the most sustainable building materials, fits well with the theme of the building. The building's 21 m long cantilever, which hovers over the water, dramatically penetrates nature. From this cantilevered space clad with transparent glass, visitors can have a unique experience of nature.

Rectangular box form wood frames of multiple modules are x-braced with steel cables to eventually create the long and slender cantilever truss structure of the Ataria Nature Interpretation Centre. By introducing thin steel cables for diagonal bracings, visual obstruction between the interior and nature in the exterior is minimized. It is not impossible to use wood for the bracing members. However, this would require much larger member sizes

Figure 3-38. Ataria Nature Interpretation Centre in Vitoria, Spain. With permission of estudio ATARIA.

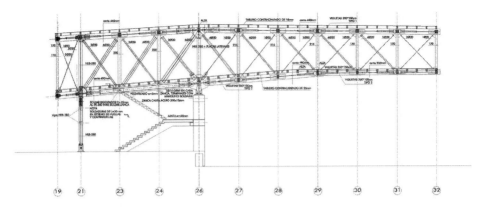

Figure 3-39. Structural drawing of the cantilever at Ataria Nature Interpretation Centre. With permission of estudio ATARIA.

and the transparency between the interior and the exterior would be much diminished.

Regarding gravity loads, the overall bending moment of the cantilever is carried by axial actions of the top and bottom chord members. The depth of the truss is larger around the support, where the overall bending moment of the cantilever is greatest, and smaller towards the free end of the cantilever, where there is no bending moment. Diagonal steel cables are also designed following the structural logic. In order to carry the gravity-induced shear forces of the cantilever by tensile actions of the cables, double or triple steel diagonal cables are placed in appropriate directions according to the required structural capacity. Not only downward gravity loads but also uplift forces by winds could also be developed. In order to carry the possible uplift force-induced shear forces also by tensile actions, the other direction diagonal steel cables are placed as well. These diagonal members are composed of single or double cables. This implies that the anticipated uplift forces are smaller than the gravity loads. These diagonal bracings in two directions eventually make the cantilever x-braced.

Diagonal bracings only in one direction instead of two directions are a possibility. In this case, however, cables cannot be used because not only tensile but also compressive forces should be carried by the diagonal members depending on either gravity or wind-induced uplift force governing cases. Therefore, instead of thin steel cables, members with greater moment of inertia, such as hollow tube or wide flange beam sections, should be considered to prevent buckling failure. This will diminish the transparent design effect of the cantilever.

There are two supports for the cantilevered truss structure. The truss is cantilevered by only about 5 percent of the entire length beyond one support (left support in Figure 3-39) and about 63 percent beyond the other support (right support in Figure 3-39). With this proportional configuration, the

Figure 3-40.
Knee bracings in the
Ataria Nature
Interpretation Centre.
With permission of
estudio ATARIA.

balancing effect of the very short left cantilever is minimal and the structure behaves like a long one-sided cantilever with a back span shorter than the length of the cantilever. In this case, under gravity loads, the right support is subjected to very large compressive force, which is even larger than the sum of the applied gravity loads, and the left support is subjected to tensile force to prevent overturning failure. Therefore, the columns and foundation for the left support should be designed to have resistance against tensile force.

Compared with the cantilevered structures presented thus far, the long cantilever of the Ataria Nature Interpretation Centre is very slender in the direction of not only gravity loads but also lateral loads. Winds can blow from any side of the cantilever. Therefore, double diagonal bracings in two directions are placed under the floor to form x-braces. For the lateral stability of the rectangular box form wood frames against wind loads from each side, internal wood knee bracings are also used.

Wozoco Apartments, Amsterdam, Netherlands

The Wozoco Apartment Building in Amsterdam designed by MVRDV is composed of 100 residential units. In order to meet the regulations about daylighting, the primary mass of the building can accommodate only 87 units facing south. In order to add the required 13 more units without violating regulations, these units are cantilevered from the north face of the main mass of the building and face east and west. This configuration resulted in an unprecedented building form.

From the large rectangular box form main mass of the building which contains 87 apartment units, five smaller masses are cantilevered only to one side. Four of the small masses are two stories tall and one is a single story tall. The length of three two-story tall cantilevers among the four is similar to the length of the back span, which is the depth of the large main building mass. With this proportion, if the cantilevers were not small portions of the main mass, the vertical end supports of the back span would be subject to tensile force development and the structure would be vulnerable to overturning failure. However, the back span of the cantilever is, in a sense, embedded into the much larger main mass of the Wozoco Apartment Building. These three two-story tall cantilevers have two, four and six floors above them. Compressive forces from these floors above the back span of the cantilever can cancel potential tensile forces developed at the vertical end supports of the back span. The remaining two- and one-story cantilevers are at the top

Figure 3-41. Wozoco Apartments in Amsterdam. Photography by Rob't Hart, image courtesy of MVRDV.

of the building. However, the lengths of these cantilevers are much shorter than the back span. This proportional configuration is not vulnerable to overturning failure.

The large cantilevers are structured with trusses. The trusses are configured in such a way that the longer diagonal members are subjected to tensile forces. The cantilevered trusses are anchored to the structurally designed demising shear walls of the residential units of the main mass. Between the units in the main building mass and the units in the cantilevered masses, there are corridors, which function as main circulation routes for the apartment building. This requires that rectangular rigid frames are inserted

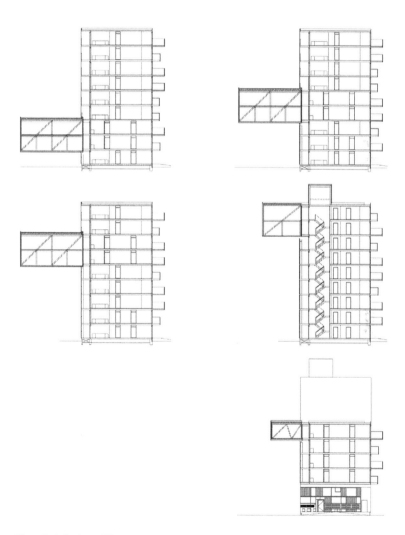

Figure 3-42. Sections of Wozoco Apartments showing five cantilevers. With permission of MVRDV.

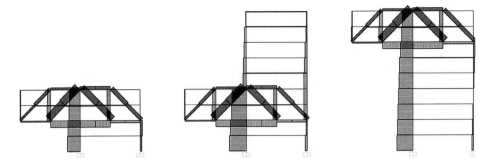

Figure 3-43. Comparative axial force diagrams of simplified Wozoco cantilevers with alternative back span structure configurations.

between the cantilevered trusses and the structural shear walls to which the cantilevers are anchored. Therefore, the loads on the cantilevered trusses are transmitted through the rectangular Vierendeel truss to the structural shear walls. Insertion of the Vierendeel truss between the triangular truss and shear wall is not a structurally efficient load carrying mechanism. However, architectural design sometimes requires this type of solution. In the Wozoco Apartment case, the inserted length of the Vierendeel trusses is very short because that is the width of the corridor. Consequently, their structural impact is relatively small.

This project introduced five cantilevers projected from the main mass of the building to solve the site specific design problems. In more general cases, this is a very expensive design solution. If possible, adding more floors, for example, to increase the number of units, will be a much more economical solution. In order to compensate for the relatively high cost of building cantilevered units, design of the other units in the main mass had to be somewhat sacrificed. The monotonously designed units in the main mass are vitalized to a certain degree to have individual identities with cantilevered balconies colored differently. The massing strategy of the 102 Dwellings in Carabanchel by Dosmasuno Arquitectos is very similar to that of the Wozoco Apartments.

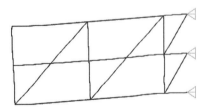
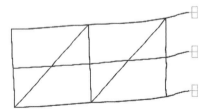

Figure 3-44. Comparative deformed shapes of simplified Wozoco cantilever trusses with and without diagonal members in the corridor.

Crane Houses in the Rheinauhafen, Cologne, Germany

Crane Houses refers to three similar looking buildings in the Rheinauhafen of Cologne, Germany, designed by Hadi Teherani Architects. Two of the three buildings are office towers and one is a residential tower. The unique form of the buildings was designed to resemble harbor cranes. Each tower is composed of two upturned L-shape volumes connected by vertical cores between them. One of the two vertical cores is located between the two vertical segments of the upturned L. The other is located between the horizontal segments of the upturned L at about a third along from the free ends.

The upturned L shape predominantly governs the visual expression of the building even with the fully exposed core between the horizontal segments of the upturned L because the core does not directly support the horizontal segment from the bottom and is clad with transparent glass. Two symmetrical deep cantilevers are projected from the exposed core in the direction perpendicular to the upturned L plane to support the horizontal segments of the upturned L. This support divides the horizontal segment of the upturned L with the back span to cantilever length ratio of about 2:1. This is close to the optimal proportioning of this type of cantilever configuration.

By using double cantilevering with the set-back vertical core, the expression of the upturned L is very dramatic as if the full length of the horizontal segment were cantilevered. With the appropriate proportioning, the horizontal segment of the upturned L including the cantilever is very

Figure 3-45. Crane Houses in the Rheinauhafen of Cologne. With permission of Rasmus Norlander.

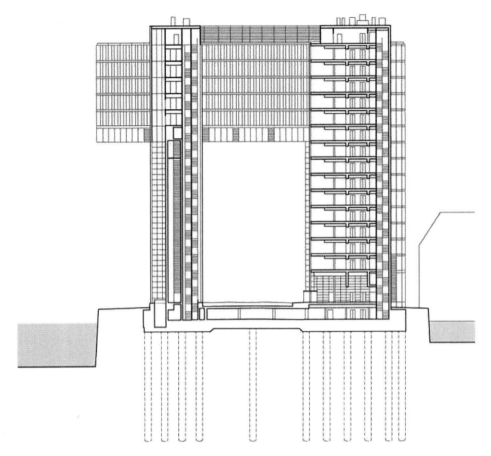

Figure 3-46. Section drawing of Crane Houses. With permission of Hadi Teherani Architects GmbH.

efficiently supported with minimized bending moments. Reinforced concrete is used as the structural material of the Crane Houses. Post-tensioning is employed for the cantilevered portions of the structure in order to better control deformations.

In the Crane Houses, the lowest levels of the horizontal segments of the upturned L are devoted as structural supports for the multiple floors above. This is clearly expressed on the building façades as thick opaque L-shaped bands. Therefore, the floors above could be architecturally designed without being much influenced by the structural challenges of the building. Alternatively, the multiple story height of the horizontal segment of the upturned L can be used as a structural depth to support the horizontal segment including the cantilever. A plausible design scenario in this case would be using large diagonal members running multiple stories to form trusses. Certainly this alternative structural design approach would influence the architectural design of the building to a greater degree.

3.4. LARGE TWO-SIDED CANTILEVERS

Compared with large one-sided cantilevers presented in the previous section, large two-sided cantilevers have greater inherent structural potential as
more efficient load carrying mechanisms. Two-sided cantilevers can be
categorized into symmetrical and asymmetrical cantilevers. While symmetrical
configurations can produce superior structural performance, their architectural expressions may not be as dynamic as asymmetrical configurations. On
the contrary, asymmetrical configurations are typically more challenging to
structure and less efficient because of their eccentricity, but their expressions
are usually more dramatic. Symmetrically configured two-sided cantilevers
naturally balance the applied gravity loads. However, in asymmetrically configured two-sided cantilevers, it is important to balance the loads if possible
especially when the common back span length between the two cantilevers
is shorter than the combined length of the two cantilevers and the degree of
asymmetry is severe.

The first diagram of Figure 3-47 shows bending moments of symmetrically configured two-sided cantilevers when the central common back
span length is 60 percent and the cantilever length is 20 percent of the entire
length of the structure. Uniformly distributed loads are applied along the beam
type structure. This proportion is close to the optimal, and the structure carries
applied loads very efficiently. As the fixed length common back span is
shifted to one side, the structure is no longer symmetrical and greater bending
moments are developed around the support of the longer cantilever (see the
second diagram of Figure 3-47). The most extreme case is produced when
the common back span is completely shifted to one side, as shown in the
third diagram of Figure 3-47. As the degree of asymmetry becomes severe,
the maximum bending moment of the structure becomes much larger than
that of the first symmetrical case. Nonetheless, no downward reaction force
is developed in any of the supports, as long as the back span length is larger
than the combined length of the two cantilevers. Downward reaction forces
involve tensile force development in the foundation system, which typically
results in a more expensive structural solution.

The first diagram of Figure 3-48 shows bending moments of symmetrically configured two-sided cantilevers when the common back span
length is 20 percent and the cantilever length is 40 percent of the entire l
ength of the structure. This is not the optimal condition structurally. Only
negative bending moments are developed throughout the structure and the
maximum bending moment value is much larger than that of the near optimal
condition shown in the first diagram of Figure 3-47. Therefore, the structure
which develops greater bending moments should be designed with heavier
structural members. However, as long as the two-sided cantilevers are
symmetrically configured, no downward reaction is developed in any of the
supports.

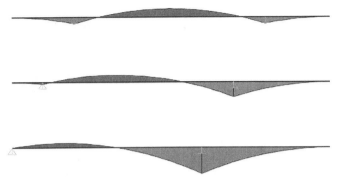

Figure 3-47.
Bending moment diagrams of symmetrically and asymmetrically configured two-sided cantilever structures with a back span longer than the combined length of the cantilevers.

Figure 3-48. Bending moment diagrams and relative support reactions of symmetrically and asymmetrically configured two-sided cantilever structures with a back span shorter than the combined length of the cantilevers.

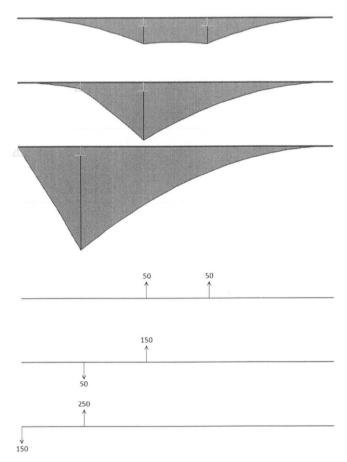

As the common back span is shifted to one side, the structure is no longer symmetrical and greater bending moments are developed around the support of the longer cantilever. The extreme case is produced when the back span is completely shifted to one side as shown in the third diagram of Figure 3-48. As the degree of asymmetry becomes severe, the maximum bending moment at the support of the longer cantilever becomes much larger than that of the first symmetrical case. The compressive upward reaction at the support of the longer cantilever becomes also very large. In addition, downward reaction force is developed at the support of the shorter cantilever, and its magnitude becomes larger as the degree of asymmetry becomes larger. This condition may require unusually large or deep foundations to prevent overturning failure of the structure. Another strategy to resolve this issue of large unbalanced load-induced overturning tendency is to apply greater loads to the shorter cantilever to balance the entire structure so the downward reaction force can be eliminated. This approach should be carefully investigated in conjunction with architectural design. For example, if the beam type structures represent simplified structural models of multistory buildings, the shorter cantilever portion can be designed with more floors than the longer cantilever portion. (See Figure 3-55, Busan Cinema Center.) Alternatively, the shorter cantilever portion can be used for functions which inherently involve heavier dead loads.

Creative Valley, Utrecht, Netherlands

The Creative Valley Building in Utrecht, Netherlands, designed by Gent & Monk Architecten is composed of ten rectangular volumes projected from a central spine structure of a narrow rectangular box form. Five volumes are projected from each of the two broader faces of the central spine structure. Among the five volumes on each side, three of them are cantilevered. The sizes of the three cantilevered volumes on each side are different. However, the sizes of each paired cantilever volumes on opposite sides are the same. Therefore, the gravity loads are balanced about the central spine structure.

Two pairs of the cantilevers are two stories tall and the remaining pair is a single story tall. The two-story tall cantilevers of the same size are placed on opposite sides of the central spine structure with a story height difference. However, the two-story tall cantilevered volumes are supported by single story tall symmetrically cantilevered trusses on the same common level of the two stories on opposite sides. Therefore, the cantilevered floors not directly supported by the symmetrically cantilevered trusses are either hung from the trusses or supported by the trusses from the bottom. By placing the cantilevered trusses symmetrically on the same level for both sides, the cantilevered volumes of the same size but different vertical locations can be efficiently supported. If the cantilevered trusses were placed at different levels to support the volumes on opposite sides, the central spine structure would be more stressed and the displacements of the cantilevers would become large.

Figure 3-49. Creative Valley Building in Utrecht, Netherlands. Photography by Abe van Ancum, image courtesy of MONK architecten.

Balancing the cantilevered masses on opposite sides of the central spine structure plays an important structural role. When only one cantilevered mass is considered on one side of the central spine without the counterbalancing mass on the other side, the structure is subjected to undesirable stresses and larger deformations. In this case, the vertical elements of the central spine closer to the cantilever are subjected to much larger compressive forces compared with the case with the balanced cantilever on the opposite side of the central spine. In addition, tensile forces are developed in the vertical elements of the central spine on the opposite side of the cantilever because the back span length of the cantilever is much smaller than the cantilevered length. Tensile forces in the vertical supports typically make the foundation system more challenging and expensive. Without the counterbalancing mass on the opposite side, the deformation of the cantilever, which directly influences serviceability of the building, becomes much larger.

Multiple cantilevered trusses are used to support each cantilevered volume in the Creative Valley Building. Therefore, the trusses are exposed to the interior space, and they may obstruct circulations and uses of the space. These exposed trusses to the interior space can be eliminated by placing only two cantilevered trusses on the perimeter of each cantilevered volume. In this case, however, much longer span floor structure of greater depth between the trusses is required. Design decisions should be made carefully considering both architectural and structural aspects holistically.

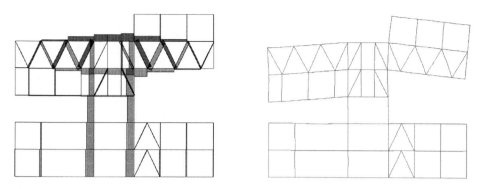

Figure 3-50. Axial force diagram and deformed shape of Creative Valley Building.

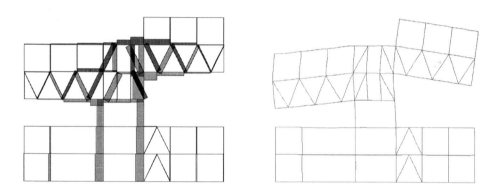

Figure 3-51. Axial force diagram and deformed shape of alternatively structured Creative Valley Building.

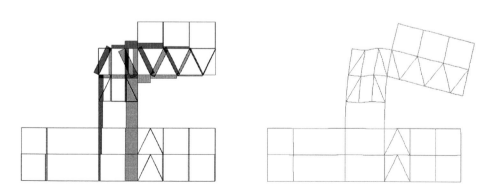

Figure 3-52. Axial force diagram and deformed shape of Creative Valley Building with a cantilever only on one side.

111

Figure 3-53. Interior view of Creative Valley Building with exposed cantilever trusses. Photography by Henny van Belkom, image courtesy of MONK architecten.

Busan Cinema Center, Busan, Korea

The Busan Cinema Center designed by Coop Himmelb(l)au is characterized by its large freeform roof structures. The project is composed of two main buildings and two unique roof (canopy) structures which cover outdoor spaces only at the top like umbrellas. The small roof structure, which measures 66 m x 100 m to 120 m, spans between the two main buildings. The big roof of 60 m x 163 m is supported by a diagrid structure of one-sheeted hyperboloid shape. In the transverse direction, this diagrid support is located around the middle and balances relatively short cantilevered roofs on opposite sides. In the longitudinal direction, the eccentric position of the diagrid support produces asymmetrically configured two-sided cantilevers of very large sizes. The longer cantilever of the big roof is an astonishing 85 m. This cantilever holds a Guinness World Record as the longest cantilever roof in the world. The shorter cantilever on the opposite side is about 50 m, which is still very long. Steel trusses are used as the structural system for the cantilevered roof structure. The depth of the trusses at the support of the longer cantilever is about 14 m, which results in a length to depth ratio of about 6. The depth of the trusses tapers towards the free end.

This roof structure in the Busan Cinema Center is not just a typical roof. This roof and the hyperboloid shape diagrid support contain substantial programmed space. The hollow hyperboloid shape structure, composed of

Figure 3-54. Busan Cinema Center. Photography by Duccio Malagamba, image courtesy of COOP HIMMELB(L)AU Wolf D. Prix & Partner.

radially finned reinforced concrete base and steel diagrids above, functions as an entrance for the complex and accommodates a café on the ground and the vertical circulations. The shorter cantilever has three levels of enclosed spaces containing restaurant, bar and lounge, which are accessible from the hyperboloid entrance structure. And both the longer and shorter cantilevers function as structural supports for the curvilinear bridges which connect the hyperboloid entrance structure and the two main buildings. The bridges are hung from the cantilevered roof structures by steel cables.

From a structural viewpoint, it is important for the shorter cantilever to contain functional spaces of three levels, which provides larger loads to

113

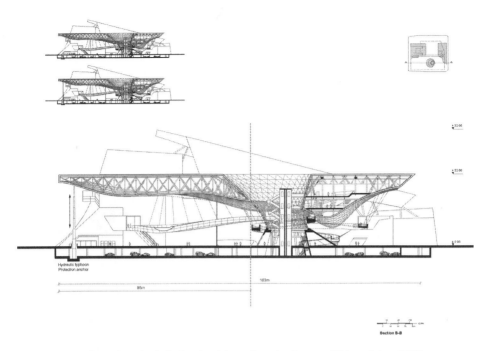

Figure 3-55. Busan Cinema Center longitudinal section of the cantilevered roof structure. With permission of COOP HIMMELB(L)AU Wolf D. Prix & Partner.

the shorter cantilever. The 85 m and 50 m cantilevers are projected from the same diagrid support in opposite directions. The structural depth of the hyperboloid diagrid support varies with the smallest of only about 20 m. With this comparatively very short common back span and asymmetrically configured large two-sided cantilevers, the roof structure is vulnerable to overturning failure with tensile force development in the diagrid support on the shorter cantilever side. The added loads to the shorter cantilever balance the overall loads of the asymmetrically configured two-sided cantilevers, and the overturning tendency is eliminated.

The Cultural Center, Castelo Branco, Portugal

The Cultural Center in Castelo Branco in Portugal designed by Josep Lluís Mateo is a bridge-like structure with large two-sided cantilevers. The Cultural Center hovering above the skate link on the ground level is supported by two reinforced concrete structures which house vertical circulations. In order to accommodate the exhibition space and auditorium, the hovering volume of the Cultural Center is significantly cantilevered in opposite directions beyond the width of the reinforced concrete core structures. The longer cantilever houses the auditorium and the shorter one contains the exhibition space. These gigantic two-sided cantilevers balance the large gravity loads applied to them

Figure 3-56. Cultural Center in Castelo Branco in Portugal. With permission of Mateo Arquitectura.

about the vertical core structure which eventually transfer the loads to the foundation system. Without the counterbalancing cantilevers, only a one-sided cantilever would make the structural design of the building more challenging.

The large cantilevers in the Cultural Center are tapered based on the functional requirements – stepped seats for the auditorium and sloped ramps for the exhibition space. These tapered forms also correspond to the structural logic of the cantilevered structure because overall bending moments in cantilever structures become larger towards the support. By tapering the cantilevered trusses, the member forces and consequently member sizes can be more equalized. In the Cultural Center, however, the member sizes of the cantilevered trusses do not quite follow this logic. In the longer cantilever containing the auditorium, bottom chord members of greater depth are used around the free end of the cantilever. This provides both an overall form definer

115

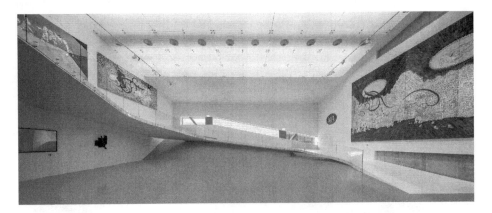

Figure 3-57. Interior view of the Cultural Center in Castelo Branco with cantilevered ramps. With permission of Mateo Arquitectura.

for the building and framing members for the sloped seats of the auditorium. Certainly, structural design does not always follow structural logic. Oftentimes it follows functionally or aesthetically determined forms.

The cantilevered trusses are geometrically configured to develop tensile forces in the relatively long diagonal members. In general, it is better to design longer members to develop tensile forces instead of compressive forces because longer members are vulnerable to buckling failure when subjected to compressive forces. If the direction of the diagonal members were reversed, they would develop compressive forces.

Inside the dramatically cantilevered Cultural Center, there is another important cantilever. The ramp in the exhibition space is cantilevered from the exterior truss wall. The reinforced concrete ramp is supported by cantilevered steel members of a tapered form. Combined with the transparent glass balustrade, the existence of the hovering cantilevered ramp is visually diminished.

The Sharp Center for Design, Ontario College of Art and Design, Toronto, Canada

The Sharp Center for Design in Toronto was designed by Alsop Architects to expand the Ontario College of Art and Design and house studios, classrooms and faculty offices. This is a very uniquely configured two story building, which hovers over and connects to the existing buildings below. The Sharp Center is basically a rectangular box of about 31 m wide 84 m long and 9 m tall, placed on top of six sets of two slanted columns of about 26 m tall and a vertical core containing elevators and a staircase. The paired slanted steel columns of a circular tube section are widely spaced on the ground but meet together at the bottom plane of the lifted rectangular box form building structure. These paired slanted columns forming triangular configurations and reinforced concrete core carry not only gravity but also lateral loads. The

Figure 3-58. Sharp Center for Design in Toronto. Photographer: Richard Johnson, Architect: Will Alsop for Alsop Architects.

columns and the core are set about 7.5 m back from the edges of the building and this produces two-sided cantilevers in both the longitudinal and transverse directions.

In the transverse direction, the two-sided cantilevers in this direction are structured with two-story tall trusses which are symmetrically supported at two points defined by two sets of the two slanted columns meeting together. Diagonal web members of these trusses beginning from the two points defined by the two sets of the two paired columns are arranged to carry the loads primarily by compressive actions. This is typically less desirable structurally. However, this configuration allows eliminating additional vertical web members in the trusses in this direction, which is advantageous for the space use because theses trusses are placed in the interior space. Since the trusses are two stories tall, the second-floor structure effectively braces the long diagonal members primarily subjected to compressions. Therefore, the vulnerability of the diagonal members to buckling failure is much reduced.

While the symmetrical configuration of the two-sided cantilevers in the transverse direction helps balancing the loads, the two-sided cantilevers in the longitudinal direction do not much influence each end cantilevers because there are very long multiple common back spans between them. The cantilevers in the longitudinal direction are also structured with two-story deep trusses on the façade planes in that direction. These longitudinal direction

117

Figure 3-59. Construction of Sharp Center for Design in Toronto. With permission of Terri Meyer Boake.

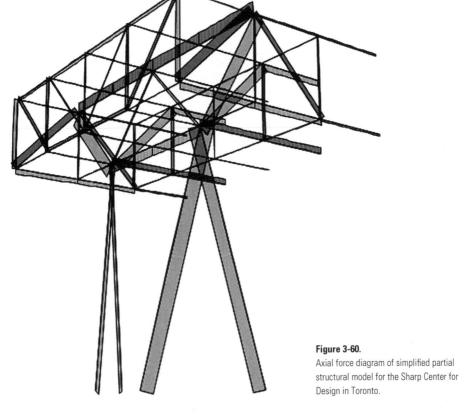

Figure 3-60.
Axial force diagram of simplified partial structural model for the Sharp Center for Design in Toronto.

trusses are supported by the free ends of the already cantilevered transverse direction trusses. Only two trusses on the longitudinal façade planes, which are spaced by about 31 m, support the cantilevers in that direction. Therefore, the two façade planes about 31 m long in the transvers direction are structured again with two-story tall trusses supported by the free ends of the cantilevered trusses in the longitudinal direction. As is the case in the Sharp Center, four-sided cantilevers are often made with this type of double cantilevering. An alternative design would be placing trusses supported by columns in both the transverse and longitudinal directions. This configuration would result in more diagonal truss members within the interior space.

The cantilevered trusses on the façades in the longitudinal direction and the long span trusses on the façades in the transverse direction are configured in such a way that the long diagonal members are primarily subjected to tensile forces. Though the large perimeter trusses are placed just behind the all façade planes, the pixelated façade design for the Sharp Center completely hides the trusses. The Statoil Oslo Office Building presented later in this chapter also uses a similar pixelated façade design. However, in the Statoil Oslo Building, trusses behind the façade are abstractly expressed through the pixilation. How to define the design relationship between the perimeter structures and building façades substantially influences not only their functional performances but also their aesthetic expressions.

3.5. MERGED CANTILEVERS

Two one-sided cantilevers are sometimes connected, typically with about 90 degrees to produce a merged cantilever. Regarding gravity loads, when the two one-sided cantilevers are identical, the structural performance of the merged cantilever is also the same as that of the individual cantilevers in term of strength and stiffness. However, when the two one-sided cantilevers are different, the resulting structural performance based on their merge is advantageous in general. When subjected to lateral loads, merged cantilevers of two one-sided cantilevers, regardless of their individual configurations, always perform better than the individual cantilevers before their merge because the merged cantilevers brace each other.

Figure 3-61 shows two identical cantilevers of the back span to cantilever length ratio of 7:3, and the merged cantilever of the two with an angle of 90 degrees. When merged, the two cantilever beams are pin-connected at their free ends in the study presented here. When subjected to uniformly distributed loads, this proportion is close to the optimal condition as has been discussed earlier, and develops minimized maximum positive and negative bending moments. The figure shows bending moment diagrams of the two cantilevers before and after merge. For clearer reading of the bending moment diagrams, the 90-degree angle between the two cantilever beams

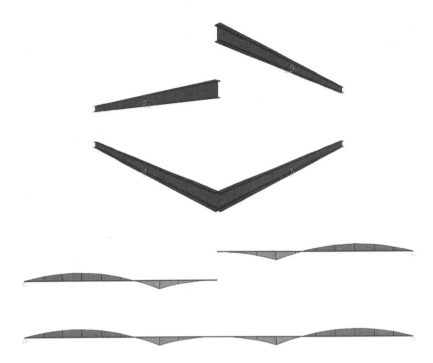

Figure 3-61. Bending moment diagrams of two identical cantilevers of a 7:3 back span to cantilever length ratio before and after merge.

was flattened. As can be seen in the figure, the bending moment diagrams of the identical individual cantilevers do not change regardless of their merge, especially when the free ends of the cantilevers are pin-connected. Consequently, regarding gravity loads, the strength and stiffness of the individual cantilevers and the merged cantilever are identical.

Figure 3-62 shows bending moment diagrams of two different one-sided cantilevers and the merged cantilever of the two. The back span to cantilever length ratios of the two cantilevered structures are 7:3 and 5:5. The total lengths of the two structures, which combine the back span and cantilever, are the same. The 7:3 cantilever is close to the optimal configuration and produces minimized peak positive and negative bending moments. The 5:5 cantilever is subjected to only negative bending moments throughout the structure. The maximum bending moment value of the 5:5 cantilever is about three times larger than that of the 7:3 cantilever. When the two structures are designed with the same structural member and subjected to identical uniformly distributed loads, the vertical displacement of the free end of the longer cantilever is about 130 times larger than that of the shorter cantilever.

When these two cantilevers are merged together with an angle of 90 degrees, the maximum negative bending moment of the shorter cantilever

120

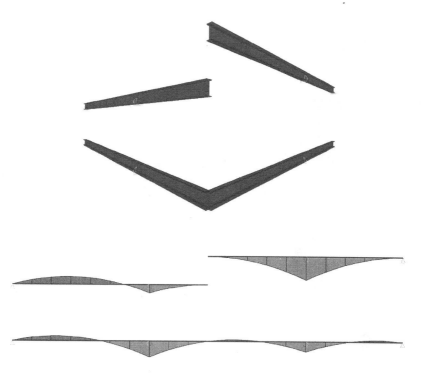

Figure 3-62. Bending moment diagrams of two cantilevers of 7:3 and 5:5 back to cantilever span length ratios before and after merge.

becomes much greater because the tip of the longer cantilever pushes down the tip of the shorter cantilever. It is as if a large downward point load is applied to the tip of the shorter cantilever. On the contrary, the maximum negative bending moment of the longer cantilever becomes much smaller because the tip of the longer cantilever is supported by the tip of the shorter cantilever. It is as if a large upward point load is applied to the tip of the longer cantilever. As a result, the bending moments of the two beams become substantially balanced by merging them. The displacement of the tip of the merged cantilever is reduced to only about 25 percent of that of the individual longer cantilever before merge.

CCTV Headquarters, Beijing, China
CCTV Headquarters in Beijing designed by OMA has introduced a new design concept of looped tall buildings. Two towers are connected at the lower and higher levels, and a complete loop is created to better satisfy the functional requirement of the building as the China Central Television Headquarters. At the top of the CCTV Headquarters, multistory-tall cantilevers projected from the gently tilted two towers meet. As in the cantilever bridges, cantilevers are used as a very efficient method of construction without falsework in the

121

Figure 3-63. CCTV headquarters in Beijing, China. CCTV/OMA Partners-in-charge: Rem Koolhaas and Ole Scheeren, designers, David Gianotten, photographed by Iwan Baan.

CCTV Headquarters. However, while the cantilevers in the typical cantilever bridges are joined and become a long span once the construction is completed, the two cantilevers in the CCTV Headquarters meet at 90 degrees and still result in an impressive merged cantilever.

The cantilever in the north–south direction slightly tapers towards the free end, which corresponds to the structural logic of cantilevers. The other cantilever, however, in the east–west direction tapers towards the support where the horizontal cantilever and the vertical tower meet. Obviously, this form reverses the structural logic. The density of the structural members is adjusted to resolve this. The structural system for the cantilevers of the building is the steel braced frame. The arrangement of the diagonal bracings becomes denser towards the supports of the cantilevers. This design approach helps efficiently increase strength and reduce deformation of the cantilevers. The lengths of the two cantilevers are very similar. With multistory-height structural depths of the cantilevers and employment of very efficient structural configurations, both cantilevers produce similar strength and stiffness. Therefore, the structural effect of merging two cantilevers in terms of carrying gravity loads is not significant in this building.

Figure 3-65 presents comparative performances of braced frame cantilever structures similar to what was employed for the CCTV Headquarters. When the diagonal bracing density is doubled throughout the cantilever, the

Figure 3-64. CCTV headquarters in Beijing, China, under construction. CCTV/OMA Partners-in-charge: Rem Koolhaas and Ole Scheeren, designers, David Gianotten.

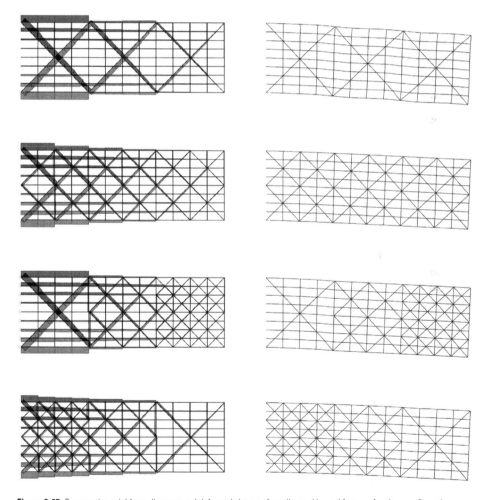

Figure 3-65. Comparative axial force diagrams and deformed shapes of cantilevered braced frames of various configurations.

123

deformation is reduced substantially (see the second case of Figure 3-65). In the original cantilever structure before increasing the density of the diagonal bracings, the figure shows that the deformation profile is clearly the result of the combined overall bending and shear actions. As the bracing density is increased, the frame-induced shear action is significantly reduced, and the overall bending action governs the deformation.

When the overall bracing members are doubled, evenly doubling bracings over the entire length of the cantilever is not the most effective strategy. As the bracing density is gradually increased from the free end to the support, the efficiency of doubled bracings can be maximized. The fourth configuration of Figure 3-65 shows an example arrangement. From the free end to one third of the cantilever, the configuration is the same as the original. From one third to two thirds of the cantilever, the bracings are doubled. From two thirds to the support of the cantilever, the bracings are quadrupled. For simple comparison, same size bracing members are used throughout the cantilever. Consequently, the overall amount of materials used for the bracings are the same for the second, third and fourth cantilevers. Among the three, the denser bracing arrangement towards the support shown in the fourth diagram produces the least amount of deformation. The deformation in this case is reduced by about 10 percent compared with the second case.

When the configuration of the gradually changing density of the bracings is reversed as shown in the third configuration of Figure 3-65, the effectiveness of doubled bracings is much reduced. Compared with the fourth case, the deformation of the third case is increased by about 20 percent. Compared with the second case, the deformation of the third case with the reversed bracing configuration produces larger deformation by about 10 percent.

When exposed, each case discussed produces quite different architectural expressions. While synergistic integration is most desirable, structural and architectural design solutions may not always go together. When they conflict, solutions should be sought carefully. For example, if the third configuration of Figure 3-65 is desired without sacrificing structural efficiency, member sizes can be adjusted accordingly. Using larger diagonal members towards the support and smaller members towards the denser free end can make this form work without diminishing structural efficiency too much. Closer collaboration between architects and engineers is essential for successful execution of building projects of complicated nature.

Nanjing Sifang Art Museum, Nanjing, China

Nanjing Sifang Art Museum designed by Steven Holl Architects is a uniquely configured three-story building. The ground and second floors are conventionally configured in this building. However, the third floor of a much differently shaped plan is significantly lifted and hovers over the second floor

Figure 3-66. Nanjing Sifang Art Museum. With permission of Steven Holl Architects.

below. The incomplete trapezoidal shaped third floor is supported by three vertical structures. Two of these structures house vertical circulation stairs and elevator each, and the third support is a planar shear wall. A fourth support would have made the structural configuration of the building less challenging, but, at the same time, not as dramatic as it is. The third floor of an incomplete trapezoidal shape supported by only three vertical structures produces multiple cantilevers including a large merged cantilever.

From the elevator tower structure and the shear wall, cantilevers are projected and they meet with an angle of about 100 degrees. This config- uration results in a large merged cantilever which defines the impressively hovering corner of the third floor. The orientation of the shear wall is taken in such a way that it can better perform not only as one of the two supports of the span between the staircase structure and the shear wall, but also as the support of the cantilever. The shear wall plane is placed in the direction of the cantilever so the large bending moment of the cantilever can be better resisted.

Steel trusses are used as the structural system for the lifted third floor including the large merged cantilever. Regarding gravity loads, the depth of the truss is the full story height, with the roof and floor edge beams as the top and bottom chord members to efficiently carry the bending moment of the cantilever. Relatively long diagonal members are placed in the directions to primarily work in tensile actions to carry the overall shear forces of the cantilever without buckling. The lengths and structural depths of the two merged cantilevers are very similar. Therefore, the effect of merging is not significant regarding gravity loads. Trusses are also placed on the roof and floor planes to carry lateral loads. The merged cantilever produces superior structural performance in terms of both strength and stiffness regarding lateral loads.

The lifted third floor is gently sloped because it mainly contains ramped gallery space. In terms of aesthetic performance of the truss structures for the lifted floating gallery space, three different types of enclosure concept are used. On the façades facing outwards, the truss structures are sandwiched by translucent enclosure materials. Therefore, in daytime, only a hint

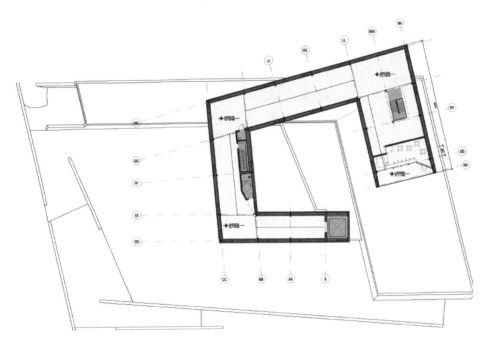

Figure 3-67. Third floor plan of Nanjing Sifang Art Museum. With permission of Steven Holl Architects.

Figure 3-68. Exterior night view and interior view of Nanjing Sifang Art Museum. With permission of Steven Holl Architects.

of trusses is expressed, and, at night, the silhouetted trusses are more clearly expressed. The inner surfaces of the exterior walls are mostly used for exhibitions, and consequently opaque enclosure materials are used and the trusses are concealed by them. The trusses on the floor are also concealed. However, the trusses on the ceiling are exposed as an interior design element. Depending on how to define the relationship between the structure and the enclosure, the aesthetic performance of the resulting design can be significantly different. Synergistic design integration should always be considered between the systems to construct built environments of higher performance.

Technological Park, Obidos, Portugal

Technological Park in Obidos, Portugal, designed by Jorge Mealha is characterized by a lifted large square donut shaped volume containing offices hovering over two long rectangular volumes on the ground with civic space and technical areas. The outer and inner square dimensions of the square donut shaped volume are about 67 m x 67 m and 59 m x 59 m, respectively. Two long rectangular volumes containing the ground floor are arranged in a V shape with an angle of about 60 degrees. The lifted square donut which is supported at six locations defines a piazza on the ground. Four of the six supports are vertical circulation cores which shoot up from the two rectangular ground floor volumes to connect them with the lifted square donut. Two more supports are added to prevent extremely long cantilevers.

The lifted square donut shaped volume is structured with trusses which are clearly expressed on the inner façade of the square donut facing the piazza. A simplified model of the square and its six supports are shown in Figure 3-71. With these support locations in relation to the lifted square, all four corners of the lifted square are merged cantilevers with an angle of 90 degrees. East and west corner merged cantilevers are composed of two equal length cantilevers. North and south corner merged cantilevers comprise two substantially different length cantilevers.

In the simplified model shown in the left diagram of Figure 3-71, the four linear elements composing the square are all connected at their ends. In the simplified model shown in the right diagram, the four linear elements composing the square are all disconnected, to comparatively evaluate structural performance of the individual and merged cantilevers. As can be seen from the bending moment diagrams, bending moments of the merged cantilevers composed of two identical cantilevers are the same as those of the individual component cantilevers before merge. However, bending moments of the merged cantilevers composed of two different cantilevers are substantially different from those of the individual component cantilevers. When merged by connecting the free ends, larger bending moments of the longer cantilever are reduced, and smaller bending moments of the shorter cantilever are increased. As a result, bending moments throughout the merged

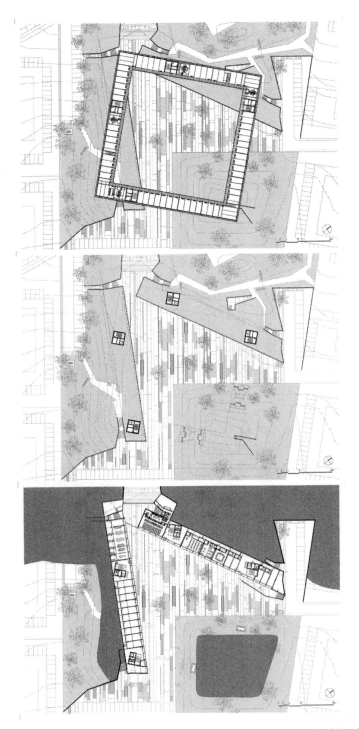

Figure 3-69. Floor plans of the Technological Park in Obidos, Portugal. With permission of Jorge Mealha Arquitectos.

Figure 3-70. Technological Park in Obidos, Portugal. With permission of João Morgado.

cantilever are substantially equalized. And the maximum bending moment of the merged structure becomes much smaller than that of the structure before merge.

In terms of displacement control, the merged cantilever is also advantageous when the component cantilevers are different. In the Technological Park in Obidos, the shorter to longer cantilever length ratio is about 2:7 for the north and south corner merged cantilevers. Compared with the displacement of the free end of the longer component cantilever, that of the merged cantilever is only about 5 percent, when the cantilevers are designed with identical structural members and the same uniformly distributed loads are applied. Certainly, the displacement of the shorter component cantilever becomes larger by about four times when it becomes a part of the merged cantilever. However, this increase is negligible compared with the decrease of the displacement of the longer component cantilever. Figure 3-71 also shows comparative deformed shapes of the simplified model of the Technological Park. In the model on the left, the square composing members are all connected, while in the model on the right, they are all disconnected. Significantly reduced vertical displacements by merging the cantilevers of two different lengths are clearly observed.

129

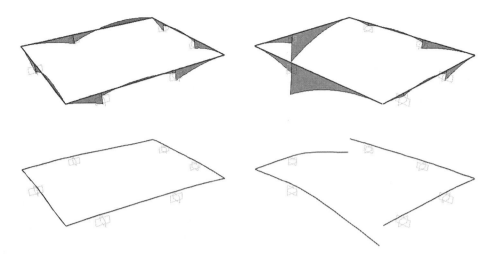

Figure 3-71. Comparative bending moment diagrams and deformed shapes of the Technological Park in Obidos with merged cantilevers (left) and independent cantilevers (right).

3.6. STACKED MULTIPLE CANTILEVERS

Multistory buildings are typically constructed by stacking the same or very similar floors vertically with elaborate vertical alignments of structural columns, floor plates and enclosures. This design and construction process produces multistory buildings very efficiently, and many mid-20th century multistory office buildings, for example, were produced by this approach. Different from this conventional approach, some of recent multistory buildings are constructed by stacking the same or different floors with intentional no vertical alignments to produce an expression of random stacking. Non-aligned stacking results in irregular cantilevers around the building.

Two different design approaches are typically used. When the desired expression of random stacking and consequently produced cantilevers are not severe, primary vertical structural members can be placed with vertical alignments as in any conventional buildings. In this case, the expression of random stacking can be produced by cantilevering horizontal structural elements such as floor beams and slabs. When the desired expression of random stacking and consequently produced cantilevers are severe, one or multiple story height module structures can be produced first and actually stacked. To support substantial cantilevers resulting from non-aligned stacking, trusses are often employed as a feasible system to frame the module structures.

With trusses, necessary façade openings can be produced through large void areas on the web portions of the trusses. When the modules structured with trusses are stacked, it is a good strategy to configure the trusses in such a way that the joints between the stacked modules coincide with the nodes of the trusses. With this type of configuration, large loads from one module to another can be transferred primarily through axial actions of the truss members. When the joints between the trussed module structures do not coincide with the nodes of the trusses, large bending moments are developed in the truss members, which should be avoided in any truss design in order to maximize the structural capacity of the system.

In stacked structures producing substantial cantilevers, it is important to carefully configure the stacking so that the resulting proportions of the back spans and cantilevers can be structurally beneficial if possible. When stacked modules create two-sided cantilevers, symmetrical configurations usually provide superior structural performance. Asymmetrical configurations could be less efficient and more vulnerable to overturning failure depending on the proportion and loading conditions, though they produce more dramatic cantilevers in general.

Halifax Library, Halifax, Canada

The Halifax Library in Halifax, Canada, by Schmidt Hammer Lassen Architects is expressed as four piled up rectangular volumes. It is, in fact, a five-story building. The second volume from the ground contains two stories – second and third floors – and the other three volumes contain one floor each. The planar dimensions of the first three volumes are very similar, while the topmost volume is much narrower with a relatively long cantilever in one direction. The second and third volumes also have some cantilevers because the shapes and planar dimensions of the stacked volumes are slightly different and their outside boundaries are not aligned to follow the different angles of the adjacent streets.

Due to this configuration, it looks like the building was constructed by randomly stacking four rectangular volumes. However, despite the randomly stacked look of the building, the primary vertical supports of this building are mostly aligned and differently cantilevered floor beams and slabs are what give it the randomly stacked expression. This is an efficient strategy to achieve this type of building form with relatively small cantilevers. The topmost volume is cantilevered much longer than the floors below. This substantial cantilever could be challenging to structure only with cantilevered floor beams and slabs. Therefore, the Vierendeel truss of the story height was used for the cantilever of the fifth floor. While the structural components up to the fourth floor are primarily constructed with reinforced concrete, the fifth-floor cantilever is mainly constructed with structural steel, which typically produces lighter structures of the same strength and stiffness compared with reinforced concrete.

As discussed earlier, the story-height Vierendeel trusses used for the cantilever carry loads primarily by bending actions of the structural members which is a very inefficient load carrying mechanism. An alternative could be using

Figure 3-72. Halifax Library. Photographer: Adam Mork, Architect: Schmidt Hammer Lassen Architects.

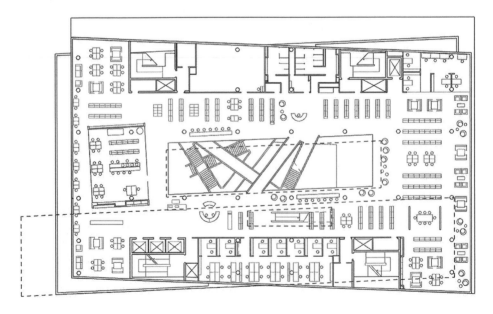

Figure 3-73. Halifax Library Floor Plan. With permission of Schmidt Hammer Lassen Architects.

triangulated normal trusses. However, this option was not employed in this building to better satisfy the functional requirements. The cantilevered portion is designed as a quiet reading room with a good view towards the harbor across the city. Diagonal members of trusses would obstruct the view, though normal trusses are a much more efficient solution than the Vierendeel trusses from a structural viewpoint. Furthermore, the cantilevered Vierendeel trusses in this building are located not on the perimeter of the cantilever but within the interior space. The floor beams supported by the cantilevered Vierendeel trusses are placed across the Vierendeel trusses and produce two symmetrical cantilevers for structural efficiency. Therefore, employing regular trusses to replace the Vierendeel trusses in this overall configuration would result in large diagonal members within the interior space.

Pierre Lassonde Pavilion at the National Museum of Fine Arts of Quebec, Canada

The expansion of the National Museum of Fine Arts of Quebec by OMA links three existing buildings, integrates the surrounding park, and tries to actively engage the city with the new building, Pierre Lassonde Pavilion. The linkage between the new and the existing buildings is made underground. The three-story above ground floors are primarily composed of stacked galleries in a cascading form. The plan dimensions of the first, second and third floor galleries are 50 m x 50 m, 45 m x 35 m and 42.5 m x 25 m, respectively. The second-floor gallery volume has a cantilever-looking 18 m portion (40 percent) out of the total length of 45 m, in the direction of the city. The third-floor gallery volume has an actual cantilever of 20 m (47 percent) out of the total length of 42.5 m, again in the direction of the city. This cascading arrangement of the masses is structurally supported by steel trusses.

The south-west elevation of the building was designed as if the second-floor volume were cantilevered from the first floor and the third floor were cantilevered from the second floor. However, in reality, the second floor is not fully cantilevered. There are vertical building core and columns under the end of the slid-out second floor volume. The vertical core and columns support the slid-out second floor with a slight setback from the south-west façade. The south-west façade design which emphasizes cascading masses, in conjunction with the set-back supports, produces an illusion of cantilever for the second-floor volume.

The third floor is actually cantilevered by slightly less than 50 percent of the entire length of the floor. If the building was designed and constructed to have an actual second floor cantilever of about 40 percent, it would be very challenging to structure the entire combined cascading cantilever of the second and third floors because the combined length and proportion of the cantilever is very large. Figure 3-78 shows comparative axial force diagrams of the two cases. When the cascading second and third floors are cantilevered, significant compressive forces are developed at the beginning of the canti-lever, and substantial tensile forces at the back spans. With only the third

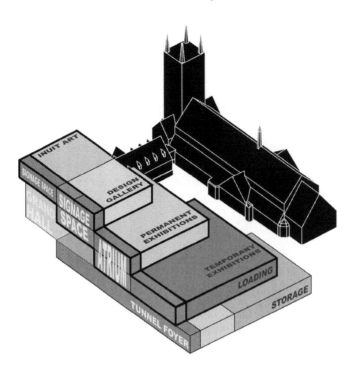

Figure 3-74. Massing diagram of the Pierre Lassonde Pavilion at the National Museum of Fine Arts of Quebec. Image courtesy of OMA.

Figure 3-75. The Pierre Lassonde Pavilion at the National Museum of Fine Arts of Quebec. Image courtesy of OMA

Figure 3-76. The Pierre Lassonde Pavilion under construction. Image courtesy of OMA.

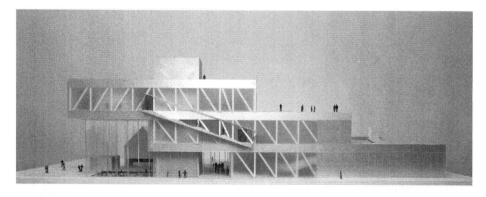

Figure 3-77. The south-west elevation view of the model of the Pierre Lassonde Pavilion at the National Museum of Fine Arts of Quebec. Image courtesy of OMA.

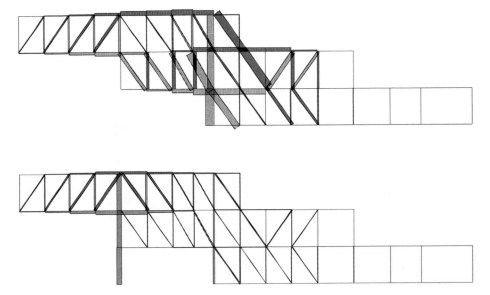

Figure 3-78. Comparative axial force diagrams of the trusses of the Pierre Lassonde Pavilion with and without vertical support under the slid-out second floor volume.

floor cantilever, the compressive force at the beginning of the cantilever is significantly reduced, and no substantial tensile forces are developed at the back spans.

Capital City Towers, Moscow, Russia

Capital City Towers in Moscow are two luxury residential buildings with a unique expression of stacked rectangular volumes. The Moscow Tower and the St. Petersburg Tower are 76 stories and 65 stories tall, respectively. The towers were designed by NBBJ and engineered by Arup. On top of the 18-story podium which contains commercial office spaces and amenities facilities for the residents in the towers, four and three rectangular volumes of about 15 stories are piled up to produce the Moscow Tower and the St. Petersburg Tower, respectively.

Though the form of the towers visually suggests that they were constructed by stacking rectangular volumes without vertical alignments, the towers were actually structured with the outrigger system composed primarily of the vertical core, outriggers, belt trusses and perimeter mega-columns. The core and perimeter mega-columns are all vertically aligned. The visual stacking effect was produced by cantilevering only slab edges. In order to give an effect of rotation to the stacked looking rectangular volumes, slab edges on only two adjacent façades of each volume are cantilevered. The slab edges on the two opposite side façades are cantilevered in the successive rectangular

Figure 3-79. Capital City Towers. With permission of Tim Griffith/NBBJ.

volume. The executed outrigger system with vertical continuity in conjunction with cantilevered slab edges is far more efficient structurally than actually stacking rectangular volumes of autonomous structures without alignment of the vertical structural elements. The concept of vertically continuous structural systems with maximized structural depth is more important in taller structures because as a building becomes taller lateral stiffness begins to govern its structural design. This subject along with more detailed discussions on the structural systems for vertical cantilevers including the outrigger system will be presented later in Chapter 4.

The façades are designed with glass curtain walls combined with a bold expression of orthogonal "super grids" of aluminum panels. The spacing of the super grids varies depending on the rectangular volume, and some of

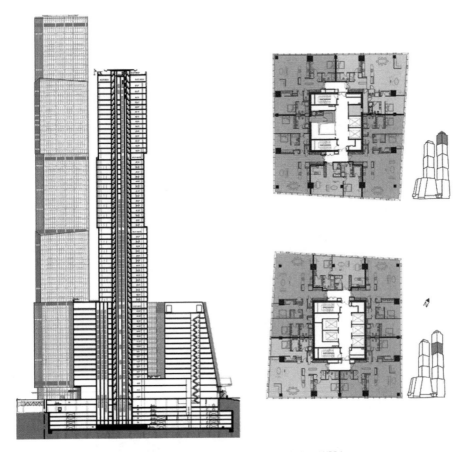

Figure 3-80. Section and typical floor plans of Capital City Towers. With permission of NBBJ.

the vertical bands of the super grids are not aligned between the rectangular volumes. This non-alignment, in combination with the expression of randomly stacked masses, produces an illusion that the buildings were actually built by stacking rectangular masses and the vertical structural elements were not aligned.

In Capital City Towers, the columns behind the cantilevered slab edges become interior columns. In today's typical tall buildings with a central core, interior columns are usually not preferred for more flexible interior spatial organizations. Depending on the function of the building, however, the level of influence of interior columns on spatial organization varies. In open office floors, interior columns could be more obtrusive, while in residential buildings, interior columns could be better integrated with necessary functional components such as demising walls.

Statoil Oslo Office Building, Oslo, Norway

The design of the Statoil Oslo Office Building by A-Lab uses the concept of actually stacking large structural modules to complete the building. Though this is not an unprecedented concept, the Statoil Oslo Office Building is one of the most recent and interesting additions to this type of architecture. The building is composed of five modules of three-story tall rectangular volume which is 140 m long and 23 m wide. The five modules are stacked in a criss-crossed form. Two modules on the ground are placed close to parallel as shown in Figure 3-81 with about a 30 m distance between them. Two other modules, again with about a 30 m distance between them, are placed on top of the two modules on the ground at an angle of about 90 degrees. The last module is placed at the top diagonally to the squarish form produced by the four modules below. According to the architect, this arrangement of the modules integrates well with the surroundings, maximizes natural light and enhances views from the building. The form created by this design concept holds a strong iconic power.

This configuration of stacking rectangular modules produces cantilevers of about 30 m maximum for the two modules in the middle tier and the last one at the top. Two three-story tall and 140 m long trusses are used on the

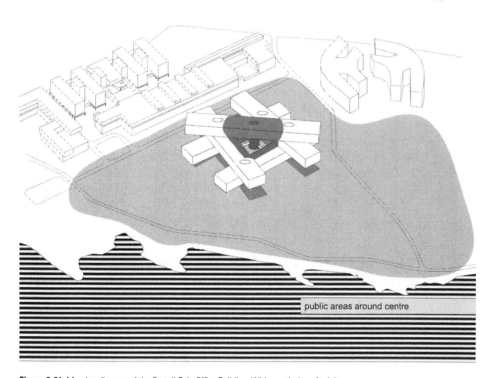

public areas around centre

Figure 3-81. Massing diagram of the Statoil Oslo Office Building. With permission of a-lab.no.

Figure 3-82. Statoil Oslo Office Building under construction showing the junctions between the rectangular volumes structured with trusses. Photo: Luis Fonseca/a-lab.no.

Figure 3-83. Statoil Oslo Office Building. Photo: Ivan Brodey/a-lab.no.

longitudinal perimeters of each module to structure it and carry the applied loads. Based on the strategic stacking of the modules, approximately maximum 20 percent of the entire length of the stacked modules is cantilevered at both ends. With this configuration, the stacked modules can be structured very efficiently with less structural materials compared with plausible alternative cases such as those with no cantilevers or much longer cantilevers.

In truss structures it is desirable whenever possible that loads are applied to the nodes instead of in the middle of the members to maximize the fundamental structural effectiveness of the system. With applied loads on the nodes, trusses carry the loads by axial actions of the members. Where the construction modules meet in this building, the geometric configurations of the web members of the trusses in the modules are adjusted so that the loads can be transferred through the nodes of the trusses.

Four vertical cores are located at the four intersecting locations of the five modules. These reinforced concrete cores, containing vertical circulations, work also as a lateral load resisting system for the building. The central area defined by the surrounding five rectangular modules is enclosed by a glass roof structure to create an atrium space. Passing through the edges of the atrium roof, it is interesting to observe that the exterior façade becomes interior atrium walls. The expression of the truss structures behind the façade is done in an abstract way by pixelating the façade.

PART II
VERTICAL CANTILEVERS

CHAPTER 4

STRUCTURAL SYSTEMS FOR TALL BUILDINGS

IN PART I OF THIS BOOK, horizontal cantilevers in buildings subjected to primarily gravity loads have been presented. Buildings are subjected to not only gravity loads but also lateral loads. With regard to lateral loads, any building should be designed as vertical cantilevers. Two primary lateral loads to be considered are wind and seismic loads. Between these two, seismic loads are more critical for low-rise buildings because low-rise buildings with high fundamental natural frequencies are much more vulnerable to the resonance conditions with the applied seismic loads of the similar frequencies. Providing more damping is a good strategy to resolve this serious structural issue. Various damping strategies for vertical cantilevers will be discussed in detail in Chapter 5. As a building becomes taller, wind loads begin to govern the structural design, and providing lateral stiffness sufficient to resist wind loads is of critical importance. This chapter presents lateral load resisting systems for tall buildings primarily against wind loads. Winds also make tall buildings laterally vibrate, and various damping strategies for wind-induced vibrations are also discussed in detail later in Chapter 5.

Tall buildings emerged in the late 19th century in the U.S. based on economic equations – increasing rentable area by stacking office spaces vertically and maximizing the rents of these offices by introducing as much natural light as possible. In traditional masonry construction, very thick and deep masonry walls were unavoidable in the lower floors of tall buildings. Heavy masonry walls with small window openings minimized the amount of daylight entering the interior spaces of early tall buildings, resulting in lower rental income. Before the invention of fluorescent lamps, daylighting was the main source of lighting in office buildings. In order to overcome this challenge

Figure 4-1. Home Insurance Building.

and serve the emerging economic driver at the time, new technologies were pursued. The result was the iron/steel skeletal structure which minimized the depth and width of structural members at the building perimeters to maximize the introduction of natural light to the interior space.

Generally, the 138 ft (42.1 m) tall 10-story Home Insurance Building of 1885 by William LeBaron Jenny in Chicago is considered as the first skyscraper. (Two more floors were added to the building in 1891 and consequently its height was increased to 180 ft (54.9 m).) This is based on the consideration of its tallness, spatial configuration related to function, and the applied technologies of the building. These factors opened a great potential for a new building type, and ultimately generated one. The combination of these criteria is of critical importance. If only the tallness of a building, which mainly contains the spaces people can occupy, is considered, some Gothic cathedrals can place the height of the Home Insurance Building underneath the vaulted ceilings of their naves. However, while a modern skyscraper has multiple stories within its height for maximum occupancy, underneath the ceiling of the nave of a Gothic cathedral is only a very high single-story space.

The importance of applied technologies in early skyscrapers exists in their potential. For instance, the heights of some early tall office buildings, such as the Montauk Building in Chicago or the Western Union Building and the Tribune Building in New York, constructed earlier than the Home Insurance Building, are comparable to, or even much greater than, that of the Home Insurance Building. Yet, they achieved their heights primarily by employing traditional load-bearing masonry structures, which required wall thicknesses of several feet on their ground levels. Thus, these earlier tall office buildings did not have the potential to grow further because of the technological limitations of their structural system. In these technological contexts in both New York and Chicago – the only two skyscraper cities in the world at that time – the invention of the iron/steel skeletal structure for the Home Insurance Building was a remarkable breakthrough towards the development of a new building type.

As a building becomes taller, the influence of lateral loads, especially wind loads, on the structural design becomes exponentially large. Eventually, for a very tall building, not strength but lateral stiffness requirement regarding wind loads is generally the governing factor of its structural design. Following the emergence of the iron/steel skeletal frame structure, various lateral load resisting systems were developed. Systems developed in the late 19th century were riveted steel connections, portal bracings and braced frames. Riveted connections were introduced in Holabird and Roche's Tacoma Building of 1889 in Chicago. Portal bracings were employed first in Burnham and Root's Monadnock Building of 1891 and Jenny's Manhattan Building of 1891 in Chicago. And braced frames were used widely. These series of structural innovations, occurring within the real estate boom in the late 1880s in Chicago, established a solid technological foundation for much taller buildings to come.

The symbolic power of skyscrapers being recognized, a notable phenomenon occurred in the development of tall buildings from the turn of the century. A skyscraper height race began, starting from the Park Row Building, which had already reached 30 stories in 1899. This height race culminated with the completion of the 102-story tall Empire State Building in 1931. Even though the heights of skyscrapers were significantly increased during this period, contrary to intuition, there had not been much conspicuous technological evolution. In terms of structural systems, most tall buildings in the early 20th century employed steel braced frames just as did those built during the previous century. Among them are the renowned Woolworth Building of 1913 and the Empire State Building. Their enormous heights at that time were accomplished not through notable technological evolution but through

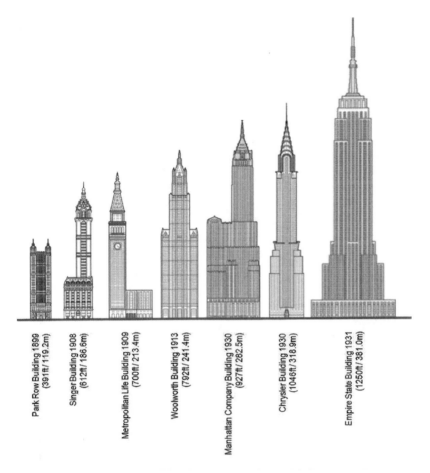

Park Row Building 1899 (391ft/119.2m)

Singer Building 1908 (612ft/186.6m)

Metropolitan Life Building 1909 (700ft/213.4m)

Woolworth Building 1913 (792ft/241.4m)

Manhattan Company Building 1930 (927ft/282.5m)

Chrysler Building 1930 (1046ft/318.9m)

Empire State Building 1931 (1250ft/381.0m)

Figure 4-2. Height race in the early 20th century.

excessive use of structural materials. Due to the absence of advanced structural analysis techniques, they were over-designed.

Structural systems for tall buildings have undergone dramatic changes in the second half of the 20th century. Such changes in the structural form and organization of tall buildings were necessitated by the emerging architectural trends in design in conjunction with the economic demands and technological developments in the realms of advanced structural analysis and design using high-speed computers. Innovative structural systems including various tube systems, outrigger systems and mixed steel-concrete composite systems are some of the new developments since the 1960s.

Since structural design of tall buildings is generally governed by lateral stiffness, one of the most important structural design considerations for tall buildings is producing higher lateral stiffness more efficiently. Tall buildings are built with an abundant amount of resources including structural materials. The amount of structural materials to carry lateral loads increases drastically as the height of the building increases. Therefore, the importance of producing higher lateral stiffness using less structural materials is of significant importance to save our limited resources and consequently create more sustainable built environments.

In order to make a tall building a more efficient vertical cantilever, it is important to maximize the structural depth of the building against lateral loads. One of the very efficient strategies to achieve this goal is to place the primary lateral load resisting components on the building perimeter. Based on the characteristic of the configuration, this type of structure can be called exterior structures. The concept of exterior structures produces superior structural solutions for tall buildings. However, placing major structural components on the building perimeter may cause limitations on the façade design. Therefore, if possible, mainly using interior structural elements, such as building cores, as the primary lateral load resisting system, is another important concept to consider. This type of system can be called interior structures. Unlike exterior structures, interior structures can provide ample flexibilities in façade design. Since the entire building depth is not fully engaged in interior structures, however, their structural capacity and efficiency may be limited. To overcome this structural limitation and, at the same time, give flexibility in façade design, a concept of interior-exterior-integrated structures, which uses both interior building core and some of the structural elements on the building perimeter, can be considered. This type of system can take advantage of the interior and exterior structures.

This chapter presents various lateral load resisting systems for tall buildings divided into these three different conceptual categories – interior structures, exterior structures and interior-exterior-integrated structures. Performance characteristics of different structural systems in each category are discussed in relation to architectural and other design-related issues theoretically and with real world examples. Furthermore, comparative performances between the systems within each category as well as between the categories are discussed.

4.1. INTERIOR STRUCTURES

Basic types of lateral load-resisting systems in the category of interior structures are the moment resisting frames, braced frames and shear walls. These systems are usually arranged as planar assemblies in two principal orthogonal directions. Moment resisting frames may be employed together with braced frames or shear walls, resulting in combined systems in which they interact.

4.1.1. Moment Resisting Frames

The moment resisting frame (MRF) typically consists of vertical columns and horizontal girders rigidly connected in a planar grid form. The overturning moment in an MRF is carried by axial actions of the columns – tension of the columns on the windward side and compression of the columns on the leeward side. However, lateral shear forces are resisted by bending of the columns and girders. Providing lateral stiffness by bending of the columns and girders is a very inefficient mechanism. This is a serious limitation of MRFs for the application to tall buildings. MRFs in vertical cantilevers subjected to lateral loads are comparable to Vierendeel trusses in horizontal cantilevers subjected to gravity loads.

Another related limiting aspect is that MRFs require progressively larger girder sizes towards the base of the building. Gravity loads in all typical floors are more or less the same in tall buildings, and, consequently, girder sizes can be very similar in all typical floors, if the girders carry only gravity loads as can be seen in the bending moment diagram of the MRF subjected to only gravity loads in Figure 4-3. However, girder sizes in MRFs need to be increased

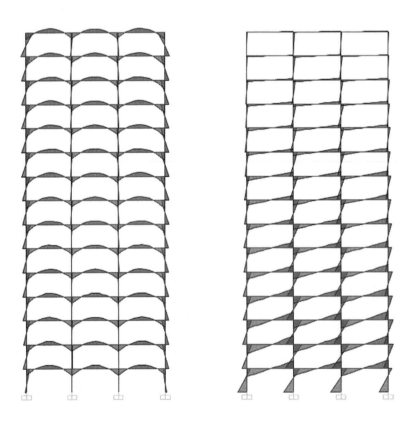

Figure 4-3. Bending moment diagrams of moment resisting frame under gravity and lateral loads.

Figure 4-4. 860 & 880 Lake Shore Drive Apartments, Chicago.

towards the base to carry the lateral loads which accumulate towards the base just like the gravity loads on columns as can be seen in the bending moment diagram of the MRF subjected to only lateral loads in the figure. This means that the floor to floor height needs to be larger in order to produce the same ceiling height for every story or the ceiling height needs to be smaller towards the base in order to keep the story height identical. Either solution may not be desirable. Due to these limitations, the maximum height of an MRF is limited to about 20–30 stories.

Figure 4-5. One Park Place, Kansas City.

For both the gravity and lateral loads, progressively larger column sizes are required towards the base of the building. The size of the columns is mainly determined by the gravity loads that accumulate towards the base of the building. Column sizes determined for the gravity loads may need to be increased to provide the required lateral stiffness of the MRF. Examples of MRFs include the 26-story tall Lake Shore Drive Apartments in Chicago designed by Ludwig Mies van der Rohe, the 19-story tall One Park Place (formerly known as Business Men's Assurance Tower) in Kansas City designed by Skidmore, Owings and Merrill, and the 27-story tall Tokyo Marine Building in Osaka designed by Kajima Design.

Tokyo Marine Building, Osaka, Japan

The Tokyo Marine Building in the Osaka business park district is a 27-story office building of steel moment resisting frame. Using the exoskeleton concept and framed columns, the moment resisting frame was uniquely employed for this building. This is a rectangular box-form building with single long spans of 21 m in the transverse direction. There are no interior columns

Figure 4-6. Tokyo Marine Building.

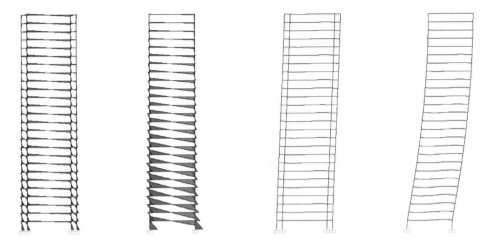

Figure 4-7. Bending moment diagram and deformed shape of the Tokyo Marine Building subjected to lateral loads in comparison with the alternatively designed frame with normal single columns instead of framed columns.

in this building and therefore column-free interior spaces are obtained through-out the building. In order to support the single long spans, the columns are composed of four members framed together with short beams of 2.7 m long, using moment connections. With these framed columns, the building's lateral stiffness is significantly improved, and the role of the floor girders as part of the lateral load resisting system is reduced to a great degree.

Figure 4-7 shows comparative bending moment and deformed shape diagrams between the simplified Tokyo Marine Building frame with the unique framed columns and an alternatively designed moment resisting frame with normal columns. When subjected to the identical lateral loads, bending moments of the moment resisting frame with the framed columns are much smaller than those of the alternatively designed normal moment resisting frame. This is because the framed columns, which are much stiffer than normal columns, carry a significant portion of the lateral loads. Consequently, deform-ation of the moment resisting frame with the framed columns employed for the Tokyo Marine Building is much smaller than that of the normal moment resisting frame.

4.1.2. Braced Frames
Braced frames resist lateral loads primarily through axial actions. The system acts as vertical cantilever trusses where the columns act as chord members and the concentric K, V, or X braces act as web members. The lateral effici-ency of the braced frame is much greater than that of the moment resisting frame because the design of the braced frame is governed by axial forces of the members while that of the moment resisting frame is governed by the bending moments of the columns and girders.

In terms of its application to tall buildings, the moment resisting frame can be applied throughout the entire building. However, the braced frame is typically applied to the building partially because the bracing members may obstruct architectural design aspects such as view, spatial organization or circulation. The building core which typically encloses vertical transportation systems such as elevators and stairwells is a common location for the braced frame.

In the braced frames, the bracings can be placed not only concentrically but also eccentrically with the nodes where typically the columns and beams meet. The former is called concentrically braced frames (CBF), and the latter, eccentrically braced frames (EBF). CBFs with triangular truss configuration provide greater lateral stiffness than EBFs. In EBFs, braces are connected to the floor girders with axial offsets. This lowers lateral stiffness but increases ductility and therefore EBFs are often used for seismic zones where ductility is an essential requirement of structural design. EBFs can also be used to accommodate wide doors or other openings, and have on occasions been used for non-seismic zones.

Figure 4-8 shows comparative axial force diagrams of single bay MRF, EBF and CBF of 10 stories subjected to lateral loads. Tension and compression in the columns resist overturning moments-induced bending in all three

Figure 4-8. Comparative axial force diagrams of MRF, EBF and CBF subjected to lateral loads (tension in darker shade and compression in lighter shade in all axial force diagrams in this chapter).

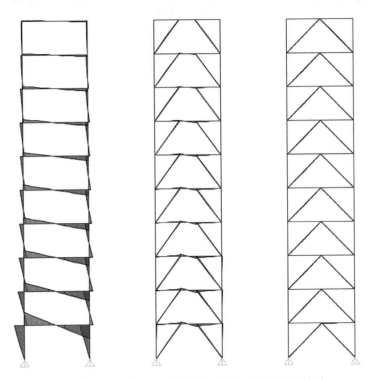

Figure 4-9. Comparative bending moment diagrams of MRF, EBF and CBF subjected to lateral loads.

Figure 4-10. Comparative deformed shapes of MRF, EBF and CBF subjected to lateral loads.

structures. Axial forces in the bracing members primarily carry lateral shear forces in the EBF and CBF. Figure 4-9 shows comparative bending moments of the same MRF, EBF and CBF subjected to lateral loads. Bending of the columns and beams carries lateral shear forces in the MRF. Relatively small bending moments are developed in the EBF, and negligible bending moments are developed in the CBF, because the EBF and CBF carry lateral shear forces primarily by axial actions of the diagonal members as shown in Figure 4-8.

Figure 4-10 shows comparative deformed shapes of the same MRF, EBF and CBF subjected to lateral loads. The lateral displacement of the MRF is significantly larger than that of the EBF or CBF, which clearly expresses inefficiency of carrying loads by bending actions of the members instead of axial actions. The EBF and CBF show superior performance in terms of lateral stiffness. Between the two, the CBF is stiffer. The stiffness of the EBF is reduced as the length of the link beam between the bracing is increased.

4.1.3. Shear Walls

Reinforced concrete shear walls have been one of the most prevalently used structural systems for tall buildings to resist lateral loads. They are treated

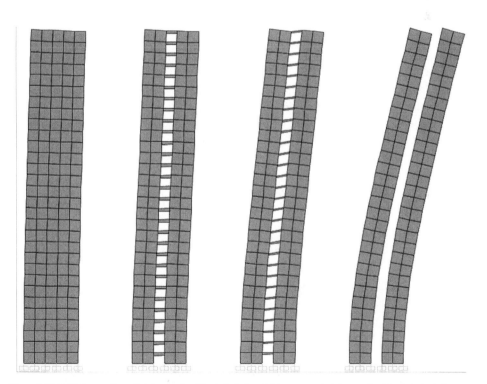

Figure 4-11. Solid shear wall, coupled shear walls with relatively deep link beams, coupled shear walls with relatively shallow link beams and two independent shear walls, all subjected to the same magnitude lateral loads.

as vertical cantilevers fixed at the base. When two or more shear walls in the same plane are interconnected by beams or slabs, the total stiffness of the system exceeds the sum of the individual wall stiffness because the connecting link beam forces the walls to act as a single unit by restraining their individual cantilever actions. These are known as coupled shear walls. Shear walls with door or window openings form coupled shear walls. The stiffness of the link beams significantly affects the performance of coupled shear walls. Stiffer link beams produce more efficient coupled shear walls. Shear walls used in tall office buildings are generally located around service and elevator cores, and stairwells. Many possibilities exist with single or multiple cores in a tall building with regard to their location, shape, number and arrangement.

Figure 4-11 shows deformed shapes of concrete shear walls of four different configurations subjected to the same magnitude lateral loads. The solid shear wall is stiffest, while the two independent shear walls are most flexible. In the coupled shear walls, the stiffness of the link beams determines the overall stiffness of the system. As the stiffness of the link beams is increased, the lateral stiffness of the coupled shear wall is also increased.

4.1.4. Shear Wall Frame Interaction System

Vertical steel trusses or reinforced concrete shear walls may be combined with rigid frames to create shear wall-frame interaction systems. Rigid frame systems are not efficient for buildings over about 30 stories in height because the lateral shear-induced deflection caused by bending of the columns and girders makes the building to sway excessively. Vertical steel shear trusses or concrete shear walls are efficient lateral load resisting systems. However, when these systems are employed partially only in building cores, the height-to-width aspect ratio of the system becomes very large, and may provide resistance for buildings up to only about 40 stories. When shear trusses or

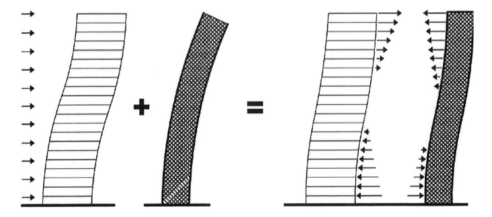

Figure 4-12. Concept diagram of shear wall-frame interaction system.

shear walls are combined with MRFs, a shear truss (or shear wall)-frame interaction system results. The upper part of the truss is restrained by the frame, whereas at the lower part, the shear wall or truss restrains the frame. This effect produces increased lateral rigidity of the building. This type of system has wide applications for buildings up to about 50 to 70 stories in height.

Figure 4-13 shows comparative axial force diagrams of 3-bay 15-story building structures of three different configurations subjected to lateral loads. The first structure is the moment resisting frame discussed in 4.1.1. In the braced hinged frame shown as the second structure in the figure, the middle bay is the braced frame studied in 4.1.2 and the two outer bays are connected to it by shear connections. Therefore, the interaction between the central braced frame and the outer bay frames is negligible. The configuration of the third structure is visually similar to the second one. However, the central braced frame and the outer bay frames are rigidly connected in the third structure. Thus, the shear truss-frame interaction is developed as has been explained.

Figure 4-14 shows comparative bending moments of the three structures. As has been discussed, lateral shear forces are carried by bending of the girders and columns in the moment resisting frame. In the braced hinged frame, negligible bending moments are developed in the frame members. This is because both overturning moments and lateral shear forces are carried primarily by axial actions of the braced frame in the middle bay and the participation of the two shear-connected outer bays in resisting lateral loads is minimal. In the shear truss-frame interaction system, lateral shear forces are carried by both axial actions of the diagonal members of the middle bay braced frame and bending of the outer bay frame members rigidly connected

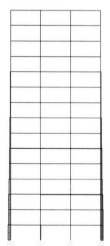
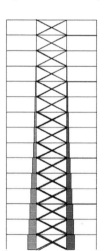
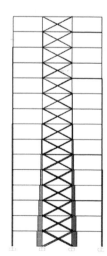

Figure 4-13. Comparative axial forces of moment resisting frame, braced hinged frame, and shear truss-frame interaction system subjected to lateral loads.

to the braced frame. Therefore, bending moments are developed in the outer bay frames as shown in the figure. However, these bending moments are much smaller than those developed in the moment resisting frame because of the greater participation of the braced frame in the middle bay in resisting lateral shear forces.

Figure 4-15 shows deformed shapes of the three structures subjected to lateral loads. The deformation of the moment resisting frame is primarily

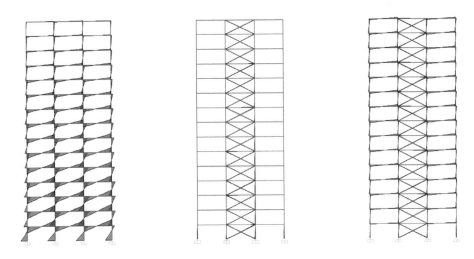

Figure 4-14. Comparative bending moments of moment resisting frame, braced hinged frame, and shear truss-frame interaction system subjected to lateral loads.

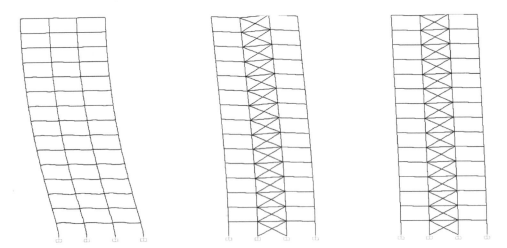

Figure 4-15. Comparative deformed shapes of moment resisting frame, braced hinged frame and shear truss-frame interaction system subjected to lateral loads.

governed by bending of the columns and girders caused by lateral shear forces. In the braced hinged frame, the deformation is mostly governed by axial actions of the central braced frame's chord members caused by overturning moment. The participation of the outer bays is negligible because they are connected to the central braced frame by shear connections. Deformation characteristics of these two frames are combined in the shear truss-frame interaction system. Large displacements of the moment resisting frames at lower levels are confined by the braced frame and those of the braced frame at higher levels are confined by the moment resisting frame. As a result, the interaction system produces superior lateral performance.

Seagram Building, New York, USA

The 38-story Seagram Building of 1958 in New York designed by Ludwig Mies van der Rohe employed shear wall-frame interaction system at lower levels,

Figure 4-16. Seagram Building. With permission of Marshall Gerometta, CTBUH (L).

shear truss-frame interaction system at mid-levels and moment resisting frame at higher levels. The shear walls were produced by embedding steel shear trusses in the concrete shear walls at lower levels of this building subjected to larger lateral loads. The shear trusses are used without concrete at the mid-levels subjected to intermediate lateral loads. And, finally, the shear trusses are eliminated at higher levels subjected to smaller lateral loads. The structural systems change logically along the height of the building. Though shear walls (or shear trusses) had widely been used in combination with moment resisting frames in tall buildings including the Seagram Building, the interaction between the systems had not fully been understood until the clear recognition of it by Fazlur Khan for the first time in the mid 1960s.

4.1.5. Staggered Truss System

The staggered truss system was developed in the mid-1960s at Massachusetts Institute of Technology as an efficient lateral load resisting system for steel tall buildings. The system is composed of story-deep trusses placed at alternate floors throughout the height of the building. The staggered trusses are also alternately placed in the longitudinal direction of the building. Typically no trusses are placed on the ground floor. Therefore, column-free interior spaces are obtained on the ground floor.

Figures 4-17 shows comparative bending moment diagrams and deformed shapes of the staggered truss system and moment resisting frame subjected to the same lateral loads. As can be seen in the figures, the major bending moments in the staggered truss system are developed only in the columns with no trusses because the alternately placed trusses carry the loads by axial actions. In the moment resisting fame, the lateral loads are carried by bending of the columns and girders, which is a very inefficient load carrying mechanism. Consequently, the moment resisting frame produces much larger deformations than the staggered truss system.

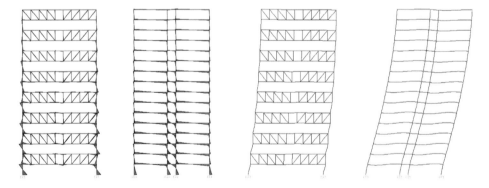

Figure 4-17. Comparative bending moment diagrams and deformed shapes of the staggered truss system and moment resisting frame subjected to the same lateral loads.

The unique compositional characteristics of the staggered truss system may impose some functional limitations to the building structured with the system. For example, column-free office spaces cannot be produced using the system. The most appropriate functions the system can accommodate are those requiring demising walls such as hotels, condos and apartments. Typically, one module of the trusses is Vierendeel without diagonals to place corridors there. It is better to place the Vierendeel module at the mid-span of the trusses because the shear force of the span is minimal there. Despite its structural efficiency, the staggered truss system is used only occasionally because of its inherent functional limitation.

4.2. EXTERIOR STRUCTURES

The nature of building perimeters has more structural significance in tall buildings than in low-rise and mid-rise buildings due to their very tallness, which means greater vulnerability to lateral loads, especially wind loads. Thus, from a structural viewpoint, it is desirable to concentrate as many lateral load-resisting system components as possible on the perimeter of tall buildings to increase their structural depth, and, in turn, their resistance to lateral loads. Typical exterior structures include tube systems, which can be defined as a fully three-dimensional structural system utilizing the entire building perimeter to resist lateral loads. Notable examples include the 100-story John Hancock Center, 83-story Aon Center, both in Chicago, and the terrorist attacked One and Two World Trade Center Towers in New York. Many other recent buildings in excess of about 50 stories have employed the tubular concept or a variation of it. Tubular structures have several types depending on their configurations and the consequent structural efficiency.

Tubular structures locate their major lateral load-resisting components at the building perimeters where building façades are, creating structural domination in the expression of the buildings. This performance-based juxtaposition naturally leads to an integrative design approach between the structural and façade systems. As a consequence, in tall buildings that employ perimeter tube type structures, technological components and architectural components of building façades are inseparable, one complementing the other. These circumstances require very intimate collaboration between architects and engineers.

4.2.1. Framed Tubes

Solely from the viewpoint of structural performance, the best tube structure is a solid tube with no fenestration. However, this type of structure cannot perform architecturally. In a framed tube system, the building has closely spaced columns and deep spandrel beams rigidly connected throughout the perimeter to minimize the area of openings. Depending on the structural geometry and proportions, exterior column spacing typically varies from about 5 to 15 ft (1.5 to 4.5 m) on centers. Practical spandrel beam depths vary from about 24 to 48 in (600 to 1200 mm).

In terms of construction, it is very time-consuming and cost-inefficient to produce a large number of rigid connections between the closely spaced perimeter columns and deep perimeter beams at the job site. In order to expedite the construction process of framed tube structures, column trees, composed of rigidly connected perimeter columns and beams of about two- to three-story tall and two- to three-bay wide, are produced at the factory and brought to the job site. Column trees are designed and produced in such a way that the connections between the column trees occur at the mid-heights of the columns and mid-spans of the beams, where bending moments are zero in the rigid frames under lateral loads because these are inflection points (see Figure 4-3).

For a framed tube subjected to lateral loads, the axial forces are greatest in the corner columns and the distribution is non-linear for both the web frames (i.e., frames parallel to wind) and the flange frames (i.e., frames perpendicular to wind). This is because the axial forces in the columns towards the middle of the flange frames lag behind those near the corner due to the bending of the spandrel beams. This phenomenon is known as

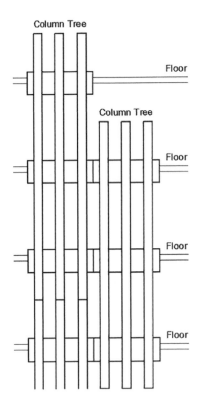

Figure 4-18.
Construction of framed tube system with column trees.

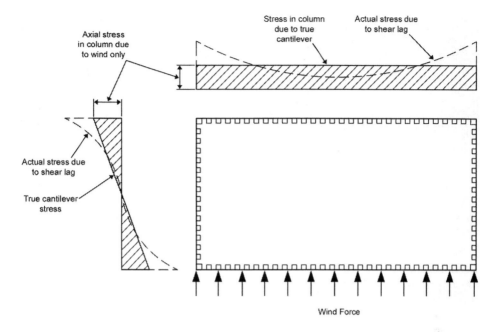

Figure 4-19. Shear lag effect in framed tube system.

shear lag. To maximize the efficiency of framed tube structures, it is important to limit the shear lag effect and aim for more cantilever-type behavior of the structure. In the braced and bundled tube structures which are discussed in more detail in the following sections, shear lag effect is reduced and structural efficiency is increased.

Figure 4-20 shows deformed shape, bending moment and axial force diagrams of a framed tube's windward flange frame. This is a 60-story framed tube model, but the diagrams show only about 10 stories from the ground for clarity. The shear lag effect is most severe at lower levels of the structure. As can be seen in the figure, the extensions of the columns are greatest at both ends, gradually reduced towards the center of the frame, and smallest at the center columns. Consequently, axial forces – tension on the windward frame in this case – are greatest at the both end columns, which also work as end columns for the web frames perpendicular to the flange frames, and gradually reduced towards the center columns. This phenomenon occurs due to the bending of the beams which rigidly connect the columns as can also be seen in the figure.

Figure 4-21 shows deformed shape, bending moment and axial force diagrams of the windward flange frame of a bundled tube composed of four identical framed tubes. Though the bundled tube system will be presented again later in this chapter, the basic compositional difference between the bundled tube and the framed tube can be noticed from the simplified plan

165

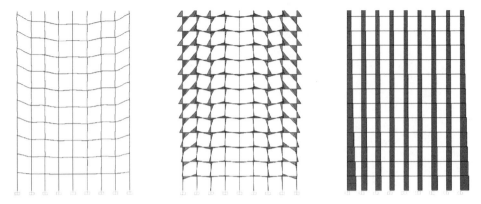

Figure 4-20. Deformed shape, bending moment and axial force diagrams of a framed tube's windward flange frame.

drawing of the bundled tube shown in the figure. The extensions of the perimeter columns are still not the same because the spandrel beams still bend and shear lag effect occurs. However, the extensions of the columns on the flange frame are much more equalized due to the additional web frame at the center across the floor plan, compared to the framed tube shown in Figure 4-20. Consequently, axial forces in the columns on the web frames are also more evenly distributed in the bundled tube than in the framed tube.

Figure 4-22 shows deformed shape, bending moment and axial force diagrams of a braced tube's windward flange frame. Though the braced tube system will also be presented again later in this chapter, the basic compositional difference between the framed tube and the braced tube can be noticed from the flange frame drawing of the braced tube shown in the figure. Lateral shear stiffness of the braced tube is much greater than that of the framed tube due to the braced web frames. Therefore, the braced tube tends to behave more like a bending beam, which is very efficient for vertical cantilevers. Though the extensions of the perimeter columns are still not the same on the flange frames, they are more equalized compared to the framed tube shown in Figure 4-20 because the diagonal members help reduce bending of the perimeter beams. Consequently, axial forces in the columns on the web frames are also more evenly distributed.

Deformation profiles of the framed tube, bundled tube and braced tube are comparatively shown in Figure 4-23. Bending of the columns and beams caused by lateral shear forces still contributes to a large degree to the deformation of the framed tube, and this tendency is reduced in the bundled tube due to its increased shear stiffness based on the added web frames. In the braced tube, with greater shear stiffness based on bracings on the web frames, axial actions of the perimeter columns govern the deformation instead of the bending of the beams and columns. Among these three, the braced tube typically produces lateral stiffness most efficiently, and the framed tube, least efficiently.

Therefore, in real world projects, the perimeter column spacing is comparatively widest in the braced tube and narrowest in the framed tube in general, though the same column spacing is used in the simplified comparative study presented here.

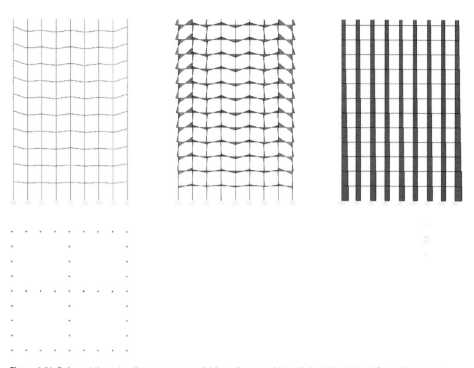

Figure 4-21. Deformed shape, bending moment and axial force diagrams of a bundled tube's windward flange frame (above) and simplified plan of the bundled tube.

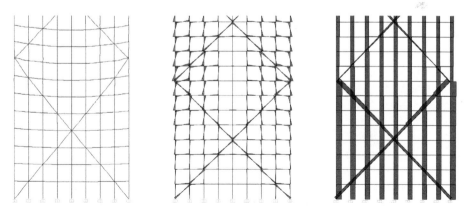

Figure 4-22. Deformed shape, bending moment and axial force diagrams of a braced tube's windward flange frame.

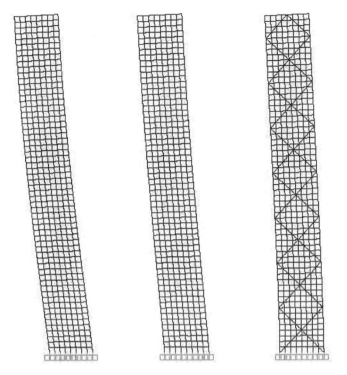

Figure 4-23. Comparative deformation profiles of the framed tube (left), bundled tube (middle) and braced tube (right) subjected to lateral loads.

One and Two World Trade Center Towers, New York (Demolished)

The World Trade Center (WTC) Tower 1 of 1972 and Tower 2 of 1973 in New York designed by the architect Minoru Yamasaki and structurally engineered by Leslie Robertson were 110-story tall twin towers. Both the 1368-ft (417 m) tall Tower 1 and 1362-ft (415.1 m) tall Tower 2 employed the framed tube system to resist lateral loads. With their plan dimensions of about 207 ft x 207 ft (63.1 m x 63.1 m), the height-to-width aspect ratio of the towers is about 6:6. The dimension of the central core is about 87 ft x 135 ft (26.5 m x 40.1 m) in the east–west direction. Therefore, the depths of the office spaces between the perimeter walls of the core and the exterior walls of the building are approximately 60 ft (18.3 m) and 36 ft (11 m) in the north–south and east–west direction respectively.

The perimeter framed tube structural system employed for the WTC Towers is typically composed of very closely spaced 14 in. (36 cm) wide columns and 52 in. (132 cm) deep spandrel beams at every floor level. The perimeter columns were made of 2.5 in. (6.4 cm) thick steel plate at lower levels and the thickness of the plate was gradually reduced to 0.25 in. (0.64 cm) at the top. The perimeter column spacing was typically 3 ft 4 in. (102 cm) on center and the typical story height was 12 ft (3.7 m).

Figure 4-24. Demolished World Trade Center in New York.

From the viewpoint of structural performance, the very narrow perimeter column spacing used in the WTC is desirable to make the structure closer to the most effective solid tube. However, this configuration in this building allowed only about 2 ft 2 in. (66 cm) wide fenestrations between the columns. With this narrow spacing, it was impossible to make entrances to the towers on the ground level having reasonably sized entrance doors. In order to resolve this issue, three columns were merged into a larger column on the ground lobby level to allow larger openings there as can be seen in Figure 4-24.

Another structural component on the perimeter of the WTC Towers was hat trusses on the roof. The hat trusses were placed between the 107th floor and roof and connected the central core structure and perimeter tube structure. The primary purpose of these hat trusses was to support antennae atop the buildings, though their configuration looks similar to the outrigger trusses in the outrigger structural systems. More detailed discussions on actual outrigger trusses, their optimal locations and performances in the outrigger structural system are presented later in this chapter.

The perimeter framed tube system, obtained by placing narrowly spaced columns and deep spandrel beams on the building perimeter, carries lateral loads applied to tall buildings very efficiently. However, the required density of the structural members on the building perimeter may seriously limit the façade design. The maximum possible vision glass area on the building perimeter to allow natural light is also significantly limited in framed tube structures. In the WTC Twin Towers in New York, the vision glass area was limited to only about 40 percent of the entire façade area.

The demolition of the towers by the terrorist attacks on September 11, 2001, produced great concern about the safety of tall buildings. Among many post-9/11 tall building design trends, the most frequently adopted one throughout the world is constructing cores of tall buildings with reinforced concrete shear walls instead of steel frames, in order to enhance performance of the core as a safer route for evacuation during emergency situations especially those involving fire and high temperature.

Aon Center, Chicago, USA

Figure 4-25. Aon Center in Chicago.

The Aon Center of 1973 in Chicago designed by Edward Durrell Stone with Perkins and Will is an 1136 ft (346.3 m) tall 83-story office tower. The perimeter framed tube was employed as the lateral load resisting system for the building. The central core of structural steel frames carries only gravity loads. The tube is composed of perimeter columns of V-shaped steel plate spaced at 10 ft (3 m) and deep channel-shaped bent plate spandrel beams.

Three-story tall column trees were shop-fabricated as construction units to expedite the construction process of the perimeter tube. Moment connections are required for the connections between the perimeter columns and spandrel beams to achieve the tubular behavior. By using column trees, these time-consuming moment connections are made in the shop under higher quality control. The column trees are connected at the job site at the mid-span of the spandrel beams with bolted connections and mid-height of the columns with welded connections at lower stories and bolted or welded connections at upper stories. The mid-span of the beams and mid-height of the columns are inflection points of the web frames of the framed tube structures subjected to lateral loads. Since there are no bending moments at inflection points, these are the best locations to make field connections of the column trees. A very similar construction mechanism was also used for the previously discussed WTC Twin Towers.

In addition to the typical framed tube structure composed of column trees, L-shaped solid steel plate columns are employed at the four corners of the building. Placing the large L-shaped columns in the building corners substantially contributes to increasing the lateral stiffness of the building. When the same amounts of structural materials are used for two tall structures shown in Figure 4-26, the structure with only four large corner columns provides larger lateral stiffness than another structure with evenly spaced smaller columns. With the perimeter framed tube combined with the four large L-shaped corner columns, the weight of the structural steel used per each square foot area of the Aon Center is 33 psf (161.1 kg/m²). The weight of the

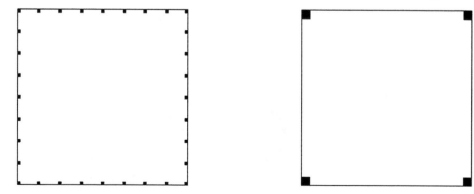

Figure 4-26. Simplified structural plans with many evenly spaced smaller perimeter columns and four large corner columns.

structural steel used for the bundled tube system, which is generally more efficient structural system than the perimeter framed tube, employed for the Willis Tower is also 33 psf. Both the Aon Center and Willis Tower are in Chicago and their height-to-width aspect ratios are about 6 and 6.4 respectively.

The V-shaped steel plate columns are integrated with the HVAC system of the building. The void spaces naturally provided by the V-shaped columns are used to contain air shafts and hot and chilled water pipes for the perimeter zone.

DeWitt-Chestnut Apartments, Chicago, USA

Figure 4-27. DeWitt-Chestnut Apartments Building in Chicago.

While the WTC Twin Towers and Aon Center are renowned supertall examples of the steel framed tube system, the framed tube concept was, in fact, employed first to the reinforced concrete DeWitt-Chestnut Apartments (now called the Plaza on DeWitt) of 1966 in Chicago designed by Skidmore, Owings and Merrill. As the very first framed tube building, the DeWitt-Chestnut Apartments is a 395 ft (120.4 m) tall 42-story building. In order to produce the tubular action, the perimeter columns are very narrowly placed at 5 ft 6 in. (1.7 m) on center. Size of the columns on the lower floors is 20 in. x 20 in. (50.8 cm x 50.8 cm) and gradually reduced to 14 in. x 14 in. (35.6 cm x 35.6 cm) at the top, as the lateral loads applied to the tower are also gradually reduced towards the top of the building. Similar to the demolished WTC Twin Towers in New York, the column spacing is increased on the ground level to accommodate more reasonably designed entrances. In order to increase the column spacing by two times, very deep transfer girders are employed where the transition occurs. And the widely spaced columns on the ground level are much larger than those above the transfer girders.

The building contains 407 apartments of studios, one-, two- and three-bedroom units. Since all lateral loads are carried by the perimeter reinforced concrete framed tube structure, great flexibility is obtained in placing interior columns. However, the exterior façade design is primarily governed by the pattern of the framed tube composed of dense structural members.

Another early example of reinforced concrete framed tube structures includes the 52-story tall One Shell Plaza of 1971 in Houston also designed by Skidmore, Owings and Merrill. In combination with the reinforced concrete shear wall core, the structural system of this building is considered as a tube-in-tube system, which is categorized as interior-exterior-integrated structures and discussed in more detail later in this chapter.

4.2.2. Braced Tubes

The framed tube becomes progressively inefficient over about 60 stories since the web frames begin to behave as conventional rigid frames. Consequently, beam and column designs are controlled by bending action, resulting in large size. Also, the cantilever behavior of the structure is thus undermined and the shear lag effect is aggravated. Figure 4-28 comparatively shows axial force and bending moment diagrams of the web frames (frames parallel to wind) of a 60-story framed tube and braced tube subjected to lateral loads.

The bending moment diagram of the framed tube represents the lateral shear force carrying mechanism of the system through bending actions of the beams and columns. The braced tube overcomes this problem by stiffening the perimeter frames in their own planes. The braced tube system can be understood as the braced frame system employed over the entire building perimeters instead of the interior core. The system is a three-dimensional vertical truss with the maximum structural depth. Consequently, lateral loads are carried very efficiently primarily by axial actions of the

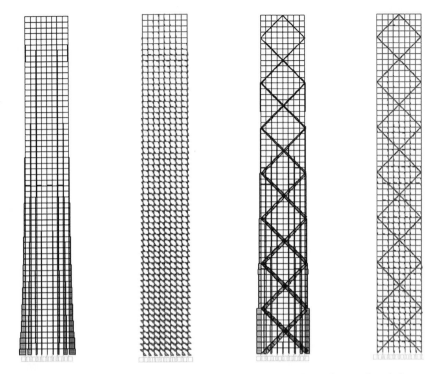

Figure 4-28. Comparative axial force and bending moment diagrams of the web frames of 60-story framed tube and braced tube.

perimeter columns and diagonal bracings. Furthermore, the diagonals of a braced tube connected to the perimeter columns at each joint effectively reduce the effects of shear lag. Therefore, the columns in the braced tube can be much more widely spaced than those in the framed tube, allowing for larger window openings, though the same column spacing is used for both the framed tube and the braced tube shown in Figure 4-28 for the purpose of direct comparison.

As can be seen in Figure 4-28, axial forces of the diagonal members represent the lateral shear force resisting mechanism of the braced tube through axial actions of the diagonal members. Bending moments of the web frame members are very small compared with those of the framed tube. Axial forces of the perimeter columns represent overturning moment resisting mechanism of the braced tube system also through axial actions.

A steel braced tube was first applied to the 100-story John Hancock Center of 1969 in Chicago. Reinforced concrete braced tube is also possible by strategically infilling the windows to create reinforced concrete bracings as is the case with the 58-story Onterie Center of 1986 in Chicago. While the braced tube is structurally very efficient, the quality of the interior space within the system may be compromised due to obstruction of the view.

In addition, construction of diagonals is by itself involved and consequently constructability could be an issue. Nonetheless, with their superior structural efficiency, braced tubes have been used for many tall buildings worldwide. Studies on the performance of braced tube structures of various configurations and more efficient design of the system are presented later in this section. Because of their enormous scale, tall buildings are built with an abundant amount of resources including structural materials. It is important to save our limited resources through efficient design to construct more sustainable built environments.

John Hancock Center, Chicago, USA

Figure 4-29. John Hancock Center in Chicago.

The John Hancock Center (now called 875 North Michigan Avenue) in Chicago designed by Skidmore, Owings and Merrill is a 1128 ft (343.7 m) tall 100-story mixed use tall building with three floors of commercial, eight floors of parking, 25 floors of office and 50 floors of condominium spaces. This is the first major tall building structured with the braced tube system. Employing the cross bracings on the building perimeter allowed the typical exterior column spacing of 40 ft (12.2 m) on the wide façade planes and 25 ft (7.6 m) on the narrow façade planes, which are incomparably larger than typical column spacing used in framed tube structures. Even with these more widely spaced perimeter columns, the braced tube system is still more efficient than the framed tube system in general. This is because the braced tube system carries lateral shear forces primarily by axial actions of the cross bracings on the web planes (planes parallel to wind) of the building, while the framed tube system carries lateral shear forces by bending actions of the perimeter columns and beams on the web planes. Apparently, carrying applied loads by axial actions is one of the most efficient load-carrying mechanisms, while carrying applied loads by bending actions is far less efficient.

The performance of the braced tube employed in the John Hancock Center is enhanced by tapering the building. The building tapers from the ground floor of about 160 ft x 260 ft (48.8 m x 79.2 m) to the roof of 100 ft x 160 ft (30.5 m x 48.8 m). This is also related to the functional requirements of the building uses on different levels. Commercial spaces do not much rely on natural light, while it is better to introduce more natural light into residential units. Therefore, deeper spaces on the lower levels accommodate commercial spaces including offices, and the spaces with relatively short clear spans between the core and exterior façades are used for residential units. More detailed discussions on the impact of tapering tall structures are presented in Chapter 6.

780 Third Avenue, New York, USA

The 780 Third Avenue of 1983 in New York also designed by Skidmore, Owings and Merrill is a 570 ft (173.7 m) tall 49-story office building. The braced tube concept of the John Hancock Center in Chicago was adopted for this building using reinforced concrete. Though the structural system is the same for the two buildings, based on the employed structural materials, construction mechanisms are different and architectural expressions are uniquely characterized accordingly. Unlike the continuous steel diagonal members across the exterior windows in the John Hancock Center, the reinforced concrete diagonals in the 780 Third Avenue were created by infilling window spaces with reinforced concrete in a diagonal pattern.

The exterior perimeter columns are spaced typically at 9 ft 4 in. (2.8 m) on center. The widths of the perimeter columns and window openings are 4 ft (1.2 m) and 5 ft 4 in. (1.6 m), respectively. The column thickness is 24 in. (61 cm) on the ground and gradually reduced to 14 in. (35.6 cm) at the top.

Figure 4-30. 780 Third Avenue in New York. With permission of John W. Cahill.

The plan dimensions of the building are 125 ft x 70 ft (38 m x 21 m). Two appropriate window openings at every level are filled on each wide face to produce X bracings, while one appropriate opening at every level is filled on the narrow face to produce single diagonal bracing in a zigzag pattern. These patterns of bracings not only produce different architectural expressions but also impact the structural performance of the system. The influence of the bracing pattern on the performance of the braced tubes is presented later in this section.

This building also has a reinforced concrete core structure composed of shear walls which houses vertical transportation. Because of the inherent characteristics of the reinforced concrete structure, the core, the exterior braced tube and flooring structures are all monolithically connected. Therefore, the systems carry lateral loads by interactions as well.

4.2.2.1 Braced Tubes of Varying Column Spacing

Braced tube structural systems are configured with perimeter diagonal bracings and vertical columns spaced evenly in general. This section investigates various perimeter column spacing strategies to improve the system's performance. Figure 4-31 shows four different cases. The braced tubes studied are 100-story tall with a story height of 3.9 m, and their plan dimensions are 54 m x 54 m. In Case 1, all the perimeter columns are spaced evenly at every 9 m. In Case 1.1, the column spacing is gradually reduced from 12 m at the mid-width of each façade plane to 6 m at the building corner, with 9 m between them. In Case 1.2, the column spacing is gradually increased from 6 m at the mid-width of each façade plane to 12 m at the building corner, with 9 m between them.

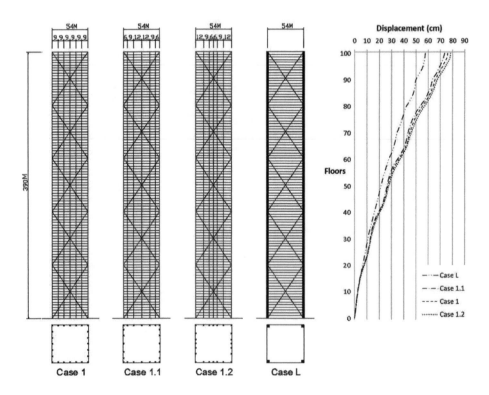

Figure 4-31. Braced tube structures configured with different column spacing strategies and their lateral displacement profiles.

The bending stiffness of each case is different because of the influence of the different column spacing, while each case's shear stiffness is almost the same regardless of the column spacing because the configuration of the bracing is unchanged. As the column spacing becomes denser towards the building corners, the contribution of the columns on the web planes (planes parallel to wind) to the bending rigidity increases, and vice versa. This phenomenon has a direct impact on the lateral displacement of each tower. When designed with the same amount of structural material, the lateral displacement of Case 1.1 is decreased and that of 1.2 is increased, compared with that of Case 1.

William LeMessurier's theoretical study of the 207-story Erewhon Center is conceptually the most extreme version of Case 1.1. The Erewhon Center uses four large corner columns in combination with X bracings between them. Regarding overturning moments, the configuration with four large corner columns produces greater bending rigidity than any first three column configurations shown in Figure 4-31, when the same quantity of structural materials is used for each alternative.

Case L of Figure 4-31 shows a 100-story braced tube structure with four large corner columns and X bracings. Except for the perimeter columns, the building's other configurations are the same as those of Case 1. While Case 1 is composed of 24 evenly spaced perimeter columns on each floor, Case L has only four large corner columns. The cross-sectional area of each column of Case L is six times larger than that of each perimeter column of Case 1 on each floor. Therefore, the quantity of structural materials used for Case 1 and L is identical. Case L, as the most extreme version of Case 1.1 conceptually, is much stiffer than Case 1.1.

Though Case L provides greater lateral stiffness based on higher bending rigidity, however, a critical design issue of this type of configuration is that it requires a far more challenging gravity load resisting system. This is because the dead and live loads of the floors must be carried also primarily by the four large corner columns spaced at 54 meters in this particular case. The maximum 54 meters cannot be simply spanned with typical deep wide flange beams. An alternative flooring system such as deep trusses should be considered.

4.2.2.2. Braced Tubes of Various Bracing Configurations

Impact of Bracing Angles

Diagonals in a braced tube structure can be configured with various different angles. Theoretically, an angle of about 35 degrees produces the maximum shear rigidity. Therefore, diagonal member sizes can be smaller as the diagonal angle becomes closer to about 35 degrees. However, smaller member sizes at each level do not guarantee the least amount of material use overall. While

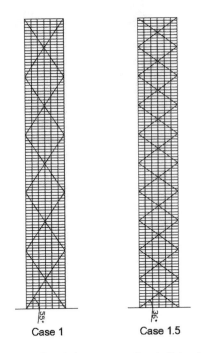

Case 1   Case 1.5

Figure 4-32. Braced tube structures configured with two different diagonal angles.

the diagonal member sizes become smaller as the angle nears 35 degrees, the total length of all diagonals decreases as the angle becomes steeper.

Figure 4-32 shows two different cases. Case 1, with diagonal bracings placed at an angle of 55 degrees, is the same structure studied in the previous section. Case 1.5 is a 100-story braced tube structure with diagonal bracings placed at 36 degrees, which is very close to the optimal angle in terms of the system's lateral shear rigidity.

Larger size diagonal members are required as the angle of diagonals deviates more from the optimal. Therefore, the required cross-sectional area of each module's diagonal members in Case 1 is much larger than that of Case 1.5 to produce the same level of lateral shear rigidity. However, the total required steel mass for the entire diagonal members of Case 1 is very similar to that of Case 1.5 because of the different total lengths of the diagonals in these two cases. Therefore, the influence of the angle of diagonal bracings on structural efficiency is minimal, if the angle is larger than the optimal within a reasonable range. In terms of constructability, Case 1.5, with the diagonal angle closer to the optimal, results in a much larger number of complicated joints than Case 1. Complicated structural joints require costly construction in general.

Impact of Bracing Shapes

While X is the most common shape of bracing for braced tubes, other types of bracings are also used. Figure 4-33 shows four different bracing types, X, chevron, alternate and single direction single diagonal bracings. Identical member sizes are used for each level bracings of Case 1 and 2, while two times larger member sizes in terms of cross-sectional area are used for those of Case 3 and 4, in order to design all these four cases with the same quantity of structural material.

The structural performance of braced tubes is influenced by the shape of bracings. Figure 4-33 also summarizes lateral deformation profiles of the four cases, based on the results analysed with structural engineering software. The case with X bracings, which are continuously connected over the entire building height, provides the greatest lateral stiffness among the four cases studied. Perimeter X bracings can be found in many tall buildings including the John Hancock Center in Chicago. Structural performances of Case 2 with chevron bracings and Case 3 with alternate direction single diagonal bracings are not much different in terms of their lateral stiffness. The Bank of the South West in Houston uses chevron bracings, and the narrow face of 780 Third Avenue in New York (Figure 4-30) uses the alternate direction

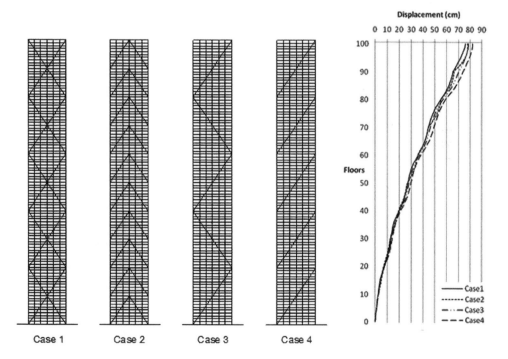

Figure 4-33. Braced tube structures with different bracing configurations and their lateral displacement profiles.

single diagonal bracings shown in Case 3. The lateral stiffness of Case 4 with single direction single diagonal bracings is smaller than that of the other three cases.

4.2.3. Bundled Tubes

Structural performance of a framed tube could be substantially improved by providing cross frames within the tube. The resulting structure could be conceived as a bundle of framed tubes. The 110-story Sears Tower (now called Willis Tower) of 1974 was the first bundled tube structure. The configuration of the bundled tube could be conceived as a large perimeter framed tube with interior web frames. The shear lag effect of the typical framed tube is substantially reduced by placing interior web frames. Therefore, the bundled tube concept allows for wider column spacing in the tubular walls, which makes it possible to place interior frame lines without seriously compromising interior space planning of the building, though the same column spacing is used for both the framed tube and the bundled tube shown earlier in Figure 4-20 and 4-21 respectively for the purpose of direct structural performance comparisons of the two systems. Due to the participation of major tube frames located inside the building by bundling multiple tubes, this system can be categorized as hybrid structures (interior-exterior integrated system) as well.

Willis Tower, Chicago, USA

The Willis Tower in Chicago designed by Skidmore, Owings and Merrill is a 1451 ft (442.1 m) tall office building. This building employs the bundled tube structure in which nine steel framed tubes are bundled at the base. Each tube is 75 ft (22.9 m) square with columns spaced at every 15 ft (4.6 m). Two diagonally positioned corner tubes are terminated at the 50th floor, and the remaining two corner tubes are terminated at the 66th floor. The cruciform bundled tube composed of five modules runs from the 67th to 90th floor. Three wings of the cruciform are dropped at the 90th floor and only two modules reach to the topmost floor. This massing strategy based on the bundled tube structural concept produced a unique stepped tapered building form much different from that typically produced by framed tube or braced tube structural concepts.

Stepped tapered form is possible typically with the introduction of large transfer girders which are very costly. In the bundled tube system employed for the Willis Tower, the stepping occurs based on the tube modules of the bundled tube. Therefore, terminating any tube module simply produces a desired step without any transfer girder. The tapered form with the greatest structural depth at the base and gradually reduced depth towards the top is structurally logical to carry lateral loads, the magnitude of which is also greatest at the base and gradually reduced towards the top.

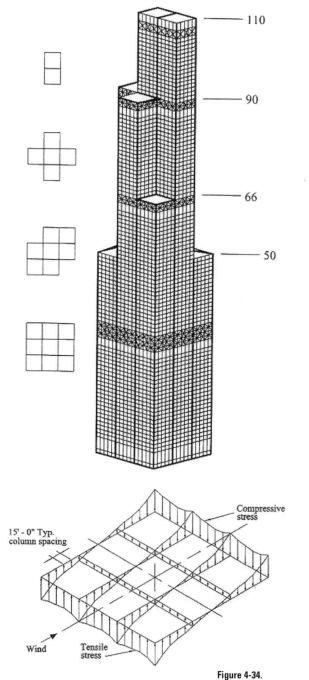

110

90

66

50

15' - 0" Typ.
column spacing

Compressive
stress

Wind

Tensile
stress

Figure 4-34.
Bundled tube systems and its improved
shear lag effect.

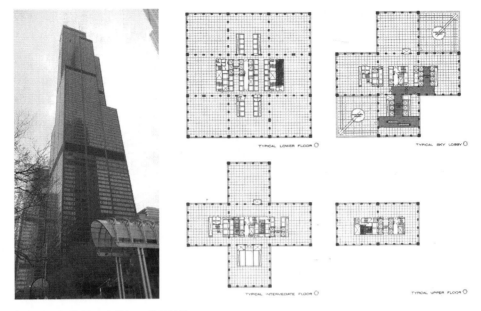

Figure 4-35. Willis Tower in Chicago. © SOM (R).

4.2.4. Diagrids

With their superior structural efficiency as a varied version of the conventional braced tube system, diagrid structures have widely been used for tall buildings recently. Early designs of tall buildings recognized the effectiveness of diagonal bracing members in resisting lateral loads. Most of the structural systems deployed for early tall buildings were steel frames with diagonal bracings of various configurations such as X, K, chevron, etc. While the structural importance of diagonals was well recognized, however, their aesthetic potential was not appreciated since they were considered to obstruct viewing the outdoors. Thus, diagonals were generally embedded within the building cores which were usually located in the interior of the building.

A major departure from this design approach occurred when braced tube structures were introduced in the late 1960s for the 100-story tall John Hancock Center in Chicago. The diagonals were located along the entire exterior perimeter surfaces of this building to maximize their structural effectiveness and capitalize on the aesthetic innovation. This strategy is much more effective than confining diagonals to narrower building cores. Despite the clear symbiosis between structural action and aesthetic intent of the Hancock Tower, this overall design approach has not emerged as the sole aesthetic preference of architects. However, recently the use of perimeter diagonals – thus the term "diagrid" – for structural effectiveness and lattice-like aesthetics has generated renewed interest in architectural and structural designers of tall buildings.

The difference between conventional braced tube structures and diagrid structures is that, for diagrid structures, almost all the conventional vertical columns are eliminated. This is possible because the diagonal members in diagrid structural systems can carry gravity as well as lateral loads due to their triangulated configuration. Compared with conventional framed tubular structures without diagonals, diagrid structures are much more effective in minimizing lateral shear deformation because they carry lateral shear by axial actions of the diagonal members on the web planes, while conventional framed tube structures carry shear by the bending of the vertical columns and horizontal spandrel beams.

Figure 4-36 shows axial force and bending moment diagrams on the web planes of a 60-story diagrid structure subjected to lateral loads. Axial forces of the diagonal members primarily represent the lateral shear force resisting mechanism through axial actions of the diagonal members. Bending moments of the diagrid members on the web planes are very small compared with those of the framed tube. Since the diagrid structure does not have conventional vertical columns, axial forces of the diagrid members on the web planes also represent the overturning moment resisting mechanism of the system, but to a minor degree. Diagrid members on the flange planes primarily participate in the overturning moment resisting mechanism.

The shear lag effect in diagrids is much smaller than that in framed tube structures because of the triangular configurations. Figure 4-37 shows deformed shape, bending moment and axial force diagrams of the diagrid structure's windward flange plane. This is a 60-story diagrid model, but the diagrams show only about 10 stories from the ground for clarity. The shear lag effect is most severe at lower levels of the structure. The extensions of the diagrid members are greatest at both ends, and gradually reduced towards the center. Consequently, axial force (tension in this case) is greatest at both ends of the diagrid members and gradually reduces towards the center diagrid members. This shear lag-induced structural inefficiency occurs due to the bending of the beams. However, compared with the conventional framed tube, the shear lag effect is smaller in the diagrid structure because the diagrids of triangular configuration are much stiffer than the framed tube of orthogonal configuration. Therefore, bending of the perimeter beams on the flange planes is much smaller in the diagrid structures, extensions of the diagrid members are more equalized, and axial forces are also more evenly distributed in the diagrid members.

When compared with braced tube structures composed of verticals and diagonals, diagrid structures carry both lateral shear forces and overturning moments by axial actions of the diagonal members, while, in the braced tube structures, the lateral shear forces are primarily carried by axial actions of the diagonals and overturning moments, by axial actions of the vertical columns. More detailed comparisons between the diagrids and braced tubes are presented later in this chapter.

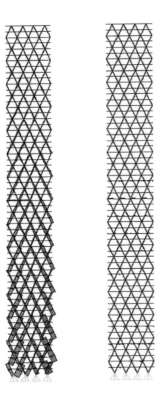

Figure 4-36. Axial force and bending moment diagrams of diagrid structure subjected to lateral loads.

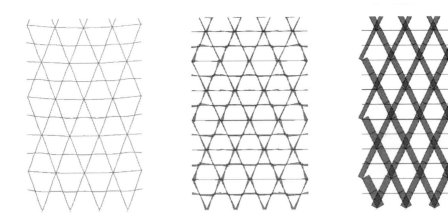

Figure 4-37. Deformed shape, bending moment and axial force diagrams of diagrid structure's windward flange frame

An early example of diagrid structures is the IBM Building of 1963 in Pittsburgh. With its 13-story building height, this building was not given much attention by architects and engineers. In the early 1980s Humana Headquarters competition, a diagrid structure was proposed by Norman Foster. However, the winning entry at that time was the post-modern style building designed by Michael Graves. Only recently have notable diagrid tall buildings been commissioned. Examples are 30 St. Mary Axe of 2003 – also known as the Swiss Re Building – in London, the Hearst Headquarters of 2006 in New York, both by Norman Foster, and Guangzhou International Finance Center of 2010 in Guangzhou, China by Wilkinson Eyre. Another important super-tall diagrid structure was proposed by Skidmore, Owings and Merrill for the Lotte Super Tower in Seoul, which employed a diagrid multi-planar façade. However, the structural design was changed from the steel diagrids to reinforced concrete outrigger structure in the newly designed Lotte World Tower by Kohn Pederson Fox. The structural concept of outrigger systems is presented later in this chapter.

The framed tube and bundled tube structures, with their dense orthogonal structural elements on the building façades, worked well with the 1960s and 1970s architecture primarily composed of verticals and horizontals. On the contrary, in contemporary urban contexts, diagrid tall structures are quite dissimilar to their tall neighbors. While many contemporary aesthetic decisions are substantially guided by subjective judgments, the use of diagrid structures stands as an innovation that requires a partnership between technical and compositional interests. These exterior structures can create a type of aesthetics, the so-called structural expression. Though the notion of structural expression is now receding with the advent of other forms of aesthetic expression at present, the diagrid system remains the exception.

Hearst Headquarters, New York, USA

The Hearst Headquarters Building in New York designed by Norman Foster is the first major diagrid tall building in North America. The 597 ft (182 m) tall 46-story building rises from the landmark façade of the existing old Hearst Building of 1926 on the site. Three sides of the building face surrounding streets – 8th Avenue, 56th Street and 57th Street. Considering this condition and better floor layout, the core is not located centrally but pushed to the side not facing the street, which reduces the benefit of the core as the main spine of the building to carry the lateral loads. With these conditions, the design evolved to finally employ the perimeter diagrid system to carry the lateral loads.

The perimeter diagrid system is used from the 10th floor to the top. Vertical mega-columns and super diagonals up to the 9th floor support the diagrid tower above. The plan dimension of the building is 120 ft x 160 ft (36.6 m x 48.8 m) with a typical story height of 14 ft (4.3 m) for the diagrid portion of the building. Three and four diamond shaped diagrid modules are placed within the 120 ft and 160 ft widths of the building, respectively. The height

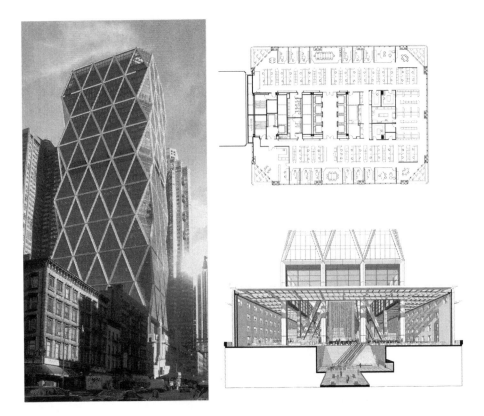

Figure 4-38. Hearst Headquarters in New York and its typical floor plan. With permission of Marshall Gerometta, CTBUH (L), Foster + Partners (RT&RB).

of the diamond shaped diagrid module is eight stories. This geometric configuration results in the diagrid angle of about 70 degrees. Since the total height of the building is about 600 ft (182.9 m), the height-to-width aspect ratio of the building is about 5. With this aspect ratio, the uniform angle of about 70 degrees is close to the optimal condition to carry the lateral loads efficiently. More detailed discussions on optimal angles of diagrid structures are presented later in this section.

When the plan dimension of 120 ft x 160 ft is used to produce a rectangular floor for every level of the diagrid tower portion of the building, cantilevers of 5, 10, 15 and maximum 20 ft (1.5, 3.0, 4.6 and 6.1 m) are repeatedly created at the four corners of the building. It is likely that the 20 ft (6.1 m) cantilever will create some concerns about floor vibration. In order to eliminate this structural issue, to produce column free corner spaces, and to create diagrid specific architectural aesthetics, corners of the Hearst Tower are chamfered following the form of the diamond shaped diagrid module. Chamfered corners also enhance aerodynamic properties of the building.

Figure 4-39. Construction of Hearst Headquarters and its prefabricated diagrid node. With permission of Michael Ficeto.

Compared with the conventional orthogonal structures, diagrid structures naturally include more complicated nodes involving six members on the diagrid planes. In the Hearst Tower, steel wide flange rolled sections are used for the diagrid members as well as for the nodes. The nodes were prefabricated and brought to the job site. The connections between the diagrid members and the nodes were done using only bolts. This method expedited the construction process significantly. If considerately designed using prefabrication strategy, constructability may not be such a limiting factor of diagrid structures.

Lotte Super Tower Project, Seoul, Korea

The 555 m tall Lotte Super Tower proposal by Skidmore, Owings and Merrill (SOM) would have become the tallest steel diagrid tower in the world.

(The final execution of the tower has been based on the design by Kohn Pederson Fox, which uses a reinforced concrete outrigger structure instead of steel diagrids.) The proposed tower possessed unique design features specific to very tall and slender diagrid structures.

The form of the SOM designed Lotte Super Tower employs an abstract regional motif coming from the shape of Chumsungdae – a celestial observatory built in 647 in Kyung-Joo, Korea. The 112-story tower has a 230 ft (70.1 m) square plan at the base which smoothly transforms to a 128 ft

Figure 4-40. Lotte Super Tower. © SOM.

(39 m) diameter circle at the top. Tapering and morphing, these are employed not only to express the desired form architecturally but also to produce a better performing building structurally. The tapered form naturally reduces wind loads applied to the tower, and the constantly changing form along the building height helps prevent formation of organized alternating vortexes around the building, which usually produce the most critical structural design condition for very tall buildings.

Diagrids are very efficient structures for very tall buildings, the design of which are primarily governed by lateral stiffness. The proposed Lotte Super Tower even further maximizes the structural potential of diagrids. The diagrids at the Lotte Super Tower are placed at different angles over the tower's height. The diagrid angles become steeper towards the ground in order to resist overturning moments more efficiently there and shallower towards the top where the impact of lateral shear forces is larger. Based on this logic, the angles of the diagrids change from about 78 degrees at the base to about 60 degrees at the top. This structural arrangement also makes visual expression of the diagrids much more dynamic. More detailed discussions on structural efficiency of varying angle diagrids are presented later in this section.

Capital Gate Tower, Abu Dhabi, UAE

The Capital Gate Tower of 2011 in Abu Dhabi designed by RMJM is an iconic freeform leaning tower in the Abu Dhabi National Exhibition Center complex and the Capital Center master development. With its 18-degree westward lean, the Capital Gate is recorded in the *Guinness Book of World Records* as the "world's furthest leaning man-made tower." Being located in the desert near the sea, the form of the tower represents a "swirling spiral of sand" and "splash" of sea water.

The Capital Gate Tower is a 164.7 m tall 36-story mixed use building with office spaces up to the 17th floor and a hotel from the 18th floor to the top. The lateral load resisting system of the Capital Gate is primarily composed of the reinforced concrete core and the perimeter diagrids. The core structure is pre-cambered by a lean of 350 mm and vertically post-tensioned to be straightened up and carry the eccentric loads from the tilted form of the building. The diagrid system with its triangular geometric configuration is very efficient to carry lateral loads by axial actions in term of structural performance and perhaps most appropriate to define the irregular freeform of the building without distortion in terms of construction. There also exist the internal diagrids in this building to form and support the hotel atrium. The perimeter diagrids are composed of steel hollow square sections of 600 x 600 mm. Because of the freeform geometry of the building, all 8,250 diagrid members are different. In order to support the building, 490 piles were driven about 20–30 meters. All of the piles were in compression at the early stage of the construction. Upon the completion of the construction, some of the piles are in tension due to the extremely tilted form of the building.

In order to enclose the perimeter diagrids of the Capital Gate Tower of irregular free form, the curtain wall units are composed of triangular glass panes following the triangular geometric configuration of the perimeter diagrids. Each two-story tall diamond shaped diagrid module is enclosed by a diamond shaped curtain wall unit composed of 18 triangular glass panes. Because of the irregular form of the building, every one of the 12,500 glass panes is different. The "splash" composed of metal meshes adds another layer on the southern façade of the office spaces to screen the sun. For the hotel, double skin façades are employed for superior environmental control.

Figure 4-41. Capital Gate Tower. With permission of Terri Boake.

O-14 Building, Dubai, UAE

While the previously presented examples are all steel diagrids, which clearly express their diagrid geometries on their façades, reinforced concrete diagrids can create concrete specific architectural aesthetic expressions quite different from those produced by steel. The O-14 Building of 2010 in Dubai by RUR Architecture employs the reinforced concrete diagrid structure as its primary lateral load-resisting system. Using the unique properties of concrete, the structural diagrid patterns, which are directly expressed as building façade aesthetics, are designed to be more fluid, irregular, and different from the

Figure 4-42. O-14 building. With permission of Reiser+Umemoto.

explicit features of steel diagrids. The reinforced concrete diagrids of the O-14 forms the exoskeleton of the 22-story office building and also function as the sun screening exterior skin of the double skin façades. The glass curtain wall façades are placed about 1 m behind the diagrid structural façades. Through the openings of the exterior diagrid skin, the 1 m deep cavity space is ventilated and heat is removed.

The diagrids are composed of perforated reinforced concrete walls, the thicknesses of which are 2 ft (61 cm) and 1 ft 3 in. (38 cm) from the ground to the 3rd level and from 3rd level to the top, respectively. Rounded square shape openings of various sizes are placed throughout the reinforced concrete exterior walls to form the unique diagrid structure. There are in total 1,326 openings, the diameter of which ranges from 4 ft 7 in. (1.4 m) to 27 ft 3 in. (8.3 m). The size, location and density of the diagrid forming openings are determined through the integrative design process to meet the architectural and structural requirements. The overall void ratio of the diagrid façades is about 45 percent in the O-14 building. The exoskeleton diagrids and interior floor slabs are connected at about 30 locations per each floor.

Different from the conventional orthogonal structures, diagrid structures involve system-specific construction challenges. In steel diagrid cases presented earlier, prefabrication technique is essential to minimize the job site work and expedite the construction process. In the O-14 building, the reinforced concrete diagrids were built using slip-form construction technique. In order to place the diagrid defining rounded square openings throughout the walls, polystyrene void forms were inserted during the placement of reinforcements.

4.2.4.1. Uniform Angle Diagrids

In diagrid tall buildings with no vertical columns, the diagonals carry not only lateral shear forces and overturning moments but also vertical gravity loads. Diagrid structures can be configured with any uniform angle to meet architectural and structural requirements. While the judgment of aesthetic expression provided by the selected angle could be a subjective matter, optimal angles for different types of loads exist in terms of structural performance. Consequently, an optimal angle for the combined loads exists for a given structure.

The optimal angle of diagonals for maximum shear rigidity for a conventional braced frame composed of vertical columns and diagonals is about 35 degrees. Overturning moments in a typical braced frame is carried by axial forces of the vertical columns, and the corresponding optimal angle of the columns is 90 degrees. In terms of gravity loads, 90-degree vertical columns are also most effective. Since the optimal angle of the columns for gravity and overturning moments is 90 degrees and that of the diagonals for maximum shear rigidity is about 35 degrees, it is expected that the optimal angle of the diagonal members in diagrid structures carrying all the combined loads will fall between these angles.

In a slender tall building design with typical maximum lateral displacement index of about a five hundredth of the building height, lateral stiffness rather than strength generally governs the structural design. Shorter buildings with low height-to-width aspect ratios behave more like shear beams, and taller buildings with high aspect ratios tend to behave more like bending beams. Thus, it is expected that as a building height is increased, the optimal diagrid angle also becomes steeper.

Figure 4-43 shows 60-story diagrid structures with a height-to-width aspect ratio of 6.5 and having various diagonal angles ranging from 53 to 76 degrees depending on the heights of diagrid modules. The building's plan dimensions are 36 m x 36 m, and its typical story height is 3.9 m. Therefore, with three diamond-shaped sub-modules placed horizontally within the building width of 36 m, 4-, 6-, 8-, 10- and 12-story tall diagrid modules result in diagrid angles of 53, 63, 69, 73 and 76 degrees, respectively. The structures are designed in such a way that lateral stiffness is provided only by the perimeter diagrids and core structures carry only gravity loads.

Each structure, assumed to be in Chicago and subjected to the code defined wind loads, is optimally designed to meet the maximum lateral displacement requirement of a five hundredth of the building height. Figure 4-43 also shows steel masses required for each structure to meet the target stiffness requirement. As can be seen from the figure, the diagrid structure configured with a diagonal angle of 69 degrees meets the design requirement with the least amount of material. As the diagrid angle is deviated from its optimal configuration, structural steel usage is increased.

Based on the similar studies with diagrid structures of various heights ranging from 40 to 100 stories, optimal angles of diagrids of different heights and height-to-width aspect ratios can be found. The structures' height-to-width aspect ratios range from about 4 for the 40-story diagrids to 10 for the 100-story diagrids. Study results show that an angle of 63 degrees is the near optimal angle for the 40- and 50-story diagrids. For the 60-story and taller diagrid structures, an angle of 69 degrees is the near optimal angle. In fact, the theoretical optimal angle for the diagrids should be increased continuously as the height-to-width aspect ratio of the building is increased. In this case, however, it is very likely that the diagrid nodes will be placed arbitrarily between the perimeter beams of two adjacent floors. This may cause some architectural, structural and constructability issues. Therefore, the diagrid module heights in this study are determined to have diagrid nodes always meeting with the perimeter beams. It should be noted that when the diagrid angle is determined to be close to the optimal with a deviation of up to about 5 degrees, structural efficiency of the system is still very high. Though the most efficient structural solution may not always best satisfy other design requirements, an integrative design approach, which considers every aspect of design holistically, should be taken to create more sustainable built environments.

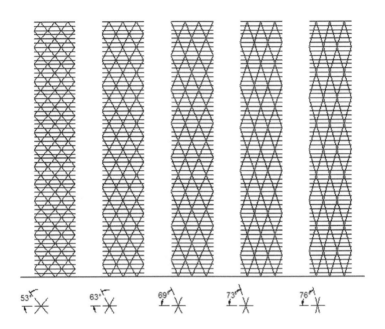

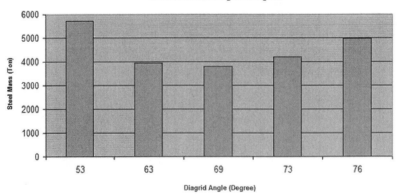

Figure 4-43. Sixty-story diagrids of various uniform angles and required structural steel for each to meet the same target stiffness requirement.

4.2.4.2. Varying Angle Diagrids

Uniform angle diagrids presented in the previous section produce very efficient structural systems for tall buildings. Structural efficiency of diagrids can be further increased for very tall buildings by varying diagrid angles. Incremental rates of shear forces and overturning moments towards the base of a tall building are different. While lateral shear forces increase almost linearly, overturning moments increase drastically towards the base of the building. Thus, in a properly designed diagrid structure, the design of the upper portion of the building is governed by shear, and the lower portion is governed by overturning moment. Considering this fact, it can be presumed that diagrid structures with gradually changing diagonal angles shall have potential for greater structural efficiency.

Comparison of the diagrids designed for the Hearst Headquarters Tower in New York and the Lotte Super Tower project in Seoul provides important structural logic related to optimal diagrid angles. Clearly, the Lotte Super Tower is a much taller building with a greater height-to-width aspect ratio. Overall, the angle of diagrids in the Lotte Super Tower is steeper than that of the Hearst Headquarters. As a building becomes taller, the optimal diagrid angle increases because a taller structure with a large height-to-width aspect ratio tends to behave more like a bending beam, and steeper angle diagonals resist overturning moments more efficiently by their axial actions. For tall diagrid structures, with height-to-width aspect ratios ranging from about 4 to 10 and diagonals placed at a uniform angle, the range of the optimal angle is from approximately 60 to 70 degrees.

It can also be noticed that the Hearst Headquarters is designed with uniform angle diagrids, while the Lotte Super Tower is designed with varying angle diagrids with steeper angles towards the base. Based on design studies, it is found that diagrid structures with diagonals placed at steeper angles towards the base generates more efficient design than those with uniform angle diagonals when the height-to-width aspect ratio of the structure is larger than about 7. However, for diagrid structures with the aspect ratio smaller than about 7, it is found that diagrid structures with uniform angle diagonals produce more efficient design. Therefore, from the viewpoint of structural engineering, it is suggested to use a varying angle diagrid structure for a very tall building with an aspect ratio greater than about 7 to save resources and, in turn, to create more sustainable built environments. Certainly, other design conditions should be carefully considered integratively to reach the final design decision.

A selection of varying angle diagrid configurations are shown as examples in Figure 4-44 for 80-story diagrid structures with a height-to-width aspect ratio of about 8. Notice that Alt. 3 is in fact a uniform angle design with the near optimal angle. Each structure is designed to meet the same maximum lateral displacement requirement of a five hundredth of the building height to find the optimal angle configuration for the diagrid of this height and height-to-width aspect ratio. As can be seen in the figure, varying angle design

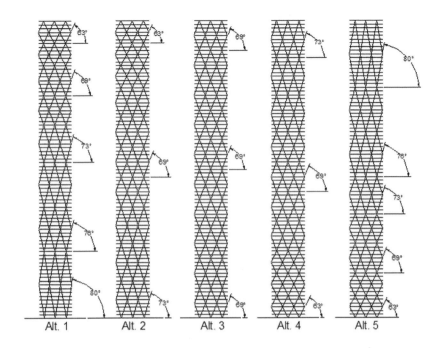

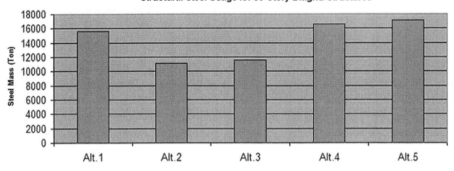

Figure 4-44. 80-story diagrids of various varying angles and required structural steel for each to meet the same target stiffness requirement.

Alt. 2, which is configured with steeper angle diagonals towards the base, uses the least amount of structural material for the 80-story diagrids. However, if the angle becomes too steep (i.e., Alt. 1), it loses its shear rigidity drastically, resulting in a less efficient solution. Alt. 4 and 5 are included for the completeness of the study, but they use more materials than other design alternatives because their grid configurations do not follow the shear and bending moment characteristics of tall buildings.

4.2.5. Space Trusses

Space truss structures, which have been used occasionally for tall buildings, are modified braced tubes with some diagonals connecting the exterior to interior. In a typical braced tube structure, all the diagonals, which connect the chord members – vertical corner columns in general, are located on the façade planes. However, in space trusses, some diagonals penetrate the interior of the building and connect the corner columns diagonally.

Bank of China Tower, Hong Kong

Figure 4-45. Bank of China Building in Hong Kong. With permission of Terri Boake.

The Bank of China Tower in Hong Kong designed by I M Pei is a 367.4 m tall 72-story office building. With the structural concept of the space truss, the building is primarily structured with four large corner columns, one large central column in the middle of the building and diagonal bracings on the building perimeter. With the five large columns, the building's floor plan and consequently mass is diagonally divided into four segments. Each triangular shape vertical mass, defined by the building's two corner columns, central column and diagonal bracings, is terminated at different heights. Therefore, the space truss structure employed for the Bank of China looks like a bundled braced tube in a sense.

Unlike the typical bundled tube structures, such as the Willis Tower in Chicago which is composed of nine bundles of framed tubes vertically cantilevered from the ground and terminating at different heights, the four triangular braced tube-like modules of the Bank of China are merged from the 25th level towards the ground. The central column is terminated at the 25th level and the loads are transferred to the four corner columns from there. This structural configuration provides a unique spatial experience as can be seen in Figure 4-46.

Figure 4-46. Bank of China interior where the central column terminates and the loads are transferred to the four corner columns.

4.3. INTERIOR-EXTERIOR INTEGRATED SYSTEMS

In the interior and exterior structures, major parts of the lateral load resisting system components are placed in the interior and on the perimeter of the building, respectively. Some structural systems for tall buildings, such as outrigger structures and tube-in-tube structures, actively engage both the interior and exterior structural components integratively to resist applied lateral loads.

4.3.1. Outrigger Structures

Compared with core shear wall structures, outrigger structures carry wind-induced overturning moments much more efficiently with greater structural depth by connecting perimeter mega-columns to stiff building cores through outriggers. The outrigger system's lateral load carrying mechanism is concep-tually explained in Figure 4-47. The overturning moment (M_o) caused by wind loads (W) is reduced due to the counteracting moment (M_c) provided by the axial actions of the perimeter mega-columns connected to the core through outriggers.

The outriggers are generally in the form of trusses in steel struc-tures or walls in reinforced concrete structures, and effectively act as stiff headers inducing a tension-compression couple in the perimeter columns. Perimeter columns connected to the outriggers are typically designed as mega-columns with very large cross-sectional area to maximize the counteracting moment by their large tension and compression forces. Belt trusses are often

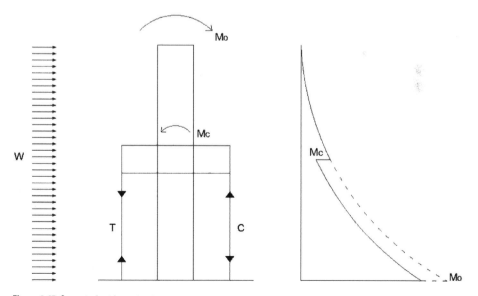

Figure 4-47. Concept of outrigger structure.

provided at the outrigger levels, especially when mega-columns are not used, to distribute the tensile and compressive forces to the exterior columns. The belt trusses also help in minimizing differential elongation and shortening of the perimeter columns.

In outrigger structures, outriggers are placed on one or multiple levels. Optimal locations of outriggers to minimize the lateral deformation have been investigated by many researchers and engineers. For the optimum performance, the outrigger in a one outrigger structure should be at about half height; the outriggers in a two outrigger structure should be at about one-third and two-thirds heights; the outriggers in a three outrigger structure should be at about one-quarter, one-half and three-quarters heights, and so on. In outrigger structures with multiple outriggers, the lowest outrigger induces the greatest resisting moment and the outriggers above carry successively less. Therefore, the outrigger at the top is least efficient and often omitted. Since the effectiveness of the outrigger structures is very much dependent upon the stiffness of the outrigger trusses, two-story tall outrigger trusses are commonly used.

Figure 4-48 shows performance of a 60-story outrigger structure with two outriggers at one-third and two-thirds heights of the building. As the vertically cantilevered braced frame core bends due to lateral loads, outrigger trusses connected to the core and the perimeter mega-columns provides resistance against the bending deformation. Curvature reversals around the outrigger truss locations shown in the deformed shape clearly show this resistance.

Architecturally, connecting the outriggers with perimeter mega-columns opens up the façade system for flexible aesthetic and architectural articulation thereby overcoming a principal drawback of closed-form tubular systems. In addition, the building's perimeter framing system may consist of simple beam-column framing without the need for more involving rigid frame type connections.

The principal disadvantages are that the outriggers interfere with the occupiable or rentable space and the lack of repetitive nature of the structural framing results in a negative impact on the erection process. However, these drawbacks can be overcome by careful architectural and structural planning such as placing outriggers in mechanical floors and development of clear erection guidelines.

The outrigger systems may be formed in any combination of steel, concrete and composite construction. Because of the many benefits of outrigger systems outlined above, this system has lately been very popular for supertall buildings all over the world. A very early example of outrigger structures can be found in the Place Victoria Office Tower (now called Stock Exchange Building) of 1964 in Montreal designed by Nervi and Moretti. It was also used by Fazlur Khan in the 42-story First Wisconsin Center of 1973 in Milwaukee, Wisconsin. However, major applications of this structural system can be seen in more recent supertall buildings such as the Jin Mao Tower of

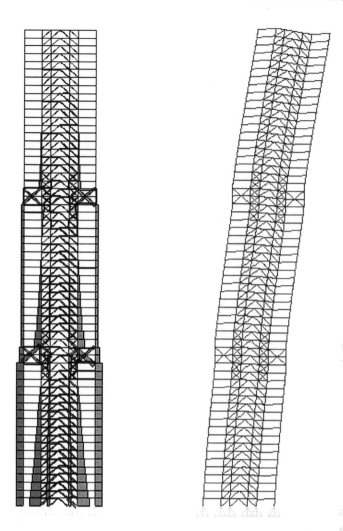

Figure 4-48. Axial force diagram and deformed shape of 60-story tall outrigger structure.

1999 in Shanghai, Taipei 101 Tower of 2004 in Taipei, International Commerce Center of 2010 in Hong Kong and Lotte World Tower of 2017 in Seoul.

Taipei 101, Taiwan

Taipei 101 designed by C.Y. Lee Architects is a 508 m tall 101-story tower with a unique profile which resembles that of Chinese old pagodas of multiple stacked units or bamboo with many joints to grow upwards with structural integrity. From the ground to the 25th floor, the tower is tapered upward with an angle of taper of about 5 degrees. Above that, eight modules of eight-story tall identical units are stacked to level 90. The units are tapered downwards with an angle of taper of 7 degrees. Finally, there are 11 mechanical levels

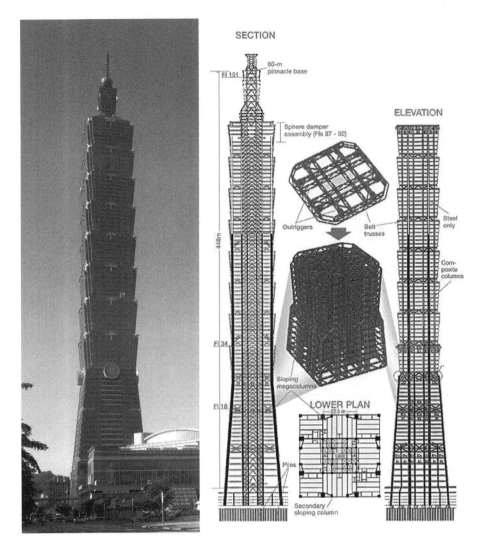

Figure 4-49. Taipei 101 and its structural concept. With permission of Thornton Tomasetti (R).

beginning from the 91st level and a 60 m tall pinnacle above that to complete the tower at the height of 508 m.

In order to resist the lateral loads, the outrigger system integrated with the architecture of the building is used. The 22.5 m x 22.5 m square shaped central core of braced frame is connected to 8 vertical mega-columns of 3 m x 2.4 m through single story height steel outrigger trusses at the lowest levels of the eight modules. These outrigger truss levels are integrated with the emergency shelter area and mechanical rooms. Double story height outrigger trusses are also located on the levels 7 and 8 and levels 17 and 18.

These outrigger trusses connect the vertical core and the sloped perimeter mega-columns from the ground to the 25th floor. Outrigger trusses at each level are connected by belt trusses.

The gravity and lateral loads are increased towards the base of the building. The geometric configuration of the outrigger system and structural material used follow this fundamental structural logic in Taipei 101. While the core structure is vertical, the mega-columns are slanted from the ground towards the 25th floor where the eight modules begin, following the profile of the building. This configuration produces the largest counteracting moment arm on the ground, where the maximum overturning moment is developed. The counteracting moment arm is gradually reduced up to the 25th floor, and constant from there to the top. The mega-columns and core columns are concrete filled steel members up to the 62nd floor and change to just steel members from there. In addition to the outrigger structural system, a pendulum type tuned mass damper (TMD), primarily composed of a 6 m diameter steel ball of 660 tonnes hung from level 92, is added to this building structure to ensure user comfort against wind induced vibration of the building. More detailed discussions on the performance of TMDs are presented in Chapter 5.

Jin Mao Tower, Shanghai, China

The Jin Mao Tower of 1999 in Shanghai designed by Skidmore, Owings and Merrill is a 420.5 m (1380 ft) tall 88-story mixed use building. The design of the ornamental tiered form was influenced by Chinese old pagodas. The eight-sided core plans and 88 stories of the tower are the result of the influence of Chinese culture which considers eight as a lucky number. Above the two-story tall main lobby, the office floors occupy the tower from the 3rd to the 50th floor. The hotel begins from the 53rd floor with a two-story tall sky lobby. The mechanical rooms are located on the 51st and 52nd floors between the office and hotel and on the penthouse floors above the hotel.

The lateral load resisting system employed for the Jin Mao Tower is the outrigger system. The system is composed of the central octagonal reinforced concrete core, eight composite perimeter mega-columns and two-story tall steel outrigger trusses which connect the core and mega-columns. The two-story tall steel outrigger trusses are located between levels 24 and 26, levels 51 and 53, and levels 85 and 87. The topmost outrigger trusses become a part of the hat truss. The outrigger trusses, except those between the levels 24 and 26, are strategically located to be integrated with the mechanical floors on the 51st and 52nd floors and the penthouse floors.

The octagonal core is normally 27 m deep and has two interconnecting web walls spaced at 9 m in two orthogonal directions on the office floors. The internal web walls are removed on the hotel floors, creating an atrium of over 30 stories. The octagonal core's perimeter wall thickness varies from 85 cm at the base to 45 cm at the top. The thickness of the interconnecting web walls

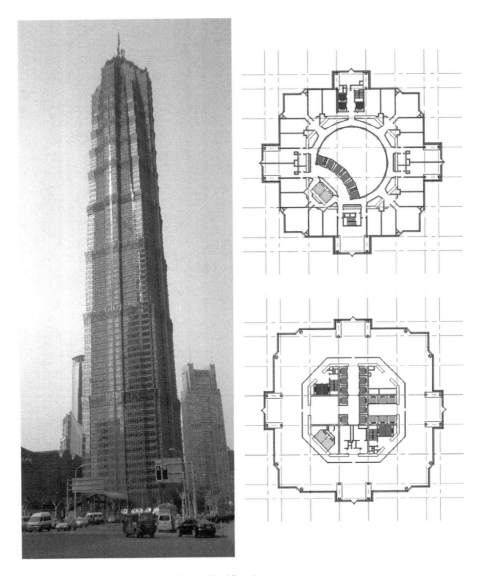

Figure 4-50. Jin Mao Tower and its typical office and hotel floor plans.

of the core is 45 cm. The steel outrigger trusses embedded in these web walls are continued from these walls and connected to the perimeter mega-columns. The size of the composite mega-columns varies form 1.5 m x 5 m at the base to 1.5 m x 3.5 m at the top.

As has been presented earlier, the performance of the outrigger system is influenced to a large degree by the vertical distributions of the outrigger trusses. Considering their structural performance, the primary outrigger trusses are located at about one third and two thirds heights of the building

Figure 4-51. Jin Mao Tower atrium.

in the Jin Mao Tower. The distance of the mega-columns from the core perimeter walls is another very important factor which determines the performance of the outrigger system. The longer, the better, as long as the outrigger trusses provide sufficient stiffness because the longer distance creates greater counteracting moment arm. In the Jin Mao Tower, the four building corner areas taper towards the top of the building. However, the central vertical strips defined by the two mega-columns spaced at 9 m on each façade do not taper in order to keep the distance between the core and the perimeter mega-columns.

4.3.2. Tube-in-Tube System

The stiffness of a perimeter tube can be increased by using the core to resist part of the lateral load, resulting in a tube-in-tube system. The floor diaphragm connecting the core and the outer tube transfers the lateral loads to both systems. The core itself could be made up of a solid tube with minimum openings, a braced tube or a framed tube. The inner tube in a tube-in-tube structure can act as a second line of defense against a malevolent attack with airplanes or missiles. For example, a solid reinforced concrete core in the World Trade Center Twin Towers in New York could probably have saved many lives of those who were trapped in fire above the levels of airplane impact. Since 9/11, reinforced concrete core is much preferred for supertall buildings.

4.4. COMPARISON BETWEEN THE SYSTEMS

The direction of the evolution of tall building structural systems – one of the most important and fundamental technological driving forces behind tall building developments – has generally been towards higher efficiency. This direction has led to today's tall building structural systems of various kinds. Among various structural systems developed for tall buildings, the systems with diagonals over the building perimeter are generally most efficient. This is because they carry lateral loads by their primary structural members' axial actions and the structural depth of the systems is maximized by placing the structural members on the building perimeter. Tall building structural systems with perimeter diagonals include braced tubes and more recently developed diagrids presented earlier in this chapter. Another very efficient structural system widely used today is outrigger structures also just presented in the previous section. Perimeter mega-columns connected to shear wall type core structure through outrigger trusses resist overturning moments very efficiently in outrigger structures. Considering their frequent employment for contemporary tall buildings due to their fundamental structural efficiency, the performance of these three structural systems is comparatively evaluated in this section.

Structural steel is used for the design of the braced tube and diagrid systems in this study. Steel is commonly used for the design and construction of these two systems in real world as well. For the outrigger system, structural steel is also used here for direct comparison with the other two systems, though, in real world, reinforced concrete and steel composite structures are very common for the outrigger system. Because of their enormous scale, tall buildings are built with an abundant amount of resources, including structural materials. More efficient structural system selection and design optimization can substantially contribute to constructing sustainable built environments by saving structural materials produced from our limited resources.

Tall buildings of 40, 60, 80 and 100 stories are designed with braced tubes, diagrids and outrigger structures in order to investigate the structural efficiency of each system comparatively depending on the building heights and height-to-width aspect ratios. The buildings' plan dimensions are 36 m x 36 m, and their typical story height is 3.9 m. Therefore, the height-to-width aspect ratios of 40- to 100-story tall buildings range from about 4 to 10. Stiffness-based design is performed for each structure to meet the target maximum allowable lateral displacement of a five hundredth of the building height. The ASCE document, Minimum Design Loads for Buildings and Other

Structures, is used to establish the wind load and the buildings are assumed to be in Chicago.

The braced tube structure's perimeter columns are spaced evenly at every 9 m, and its diagonals run 10 stories. This diagonal configuration creates an angle of 47.3 degrees measured from the horizontal. All the required lateral stiffness of the braced tube is allocated to the perimeter braced tube, and, consequently, the core structure is designed to carry only gravity loads in this study. The diagrid structure is configured with diamond-shaped sub-modules. Three diamond shaped sub-modules, the height of which is 8 stories, fit within the building width. This geometric configuration results in the diagrid angle of 69 degrees, which is very close to the optimal condition. All the required lateral stiffness of the diagrid structure is also allocated to the perimeter diagrids, and the core structure is designed to carry only gravity loads. In the outrigger structure, the core structure is a steel braced frame, which carries not only gravity but also lateral loads. Plan dimensions of the central core are 18 m x 18 m. Two-story tall outrigger trusses, which connect the braced core and perimeter mega-columns, are located at every 20 floors except at the top. Therefore, for example, outrigger trusses are located at one third and two thirds heights of the building for the 60-story outrigger structure as can be seen in Figure 4-52.

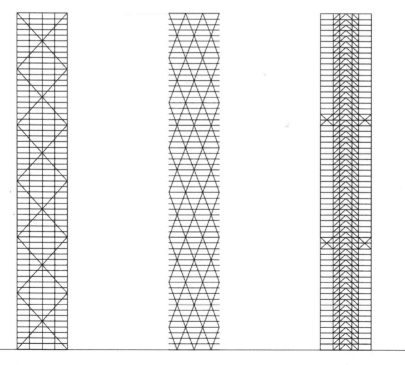

Figure 4-52. Comparative study models of braced tube, diagrids and outrigger structure.

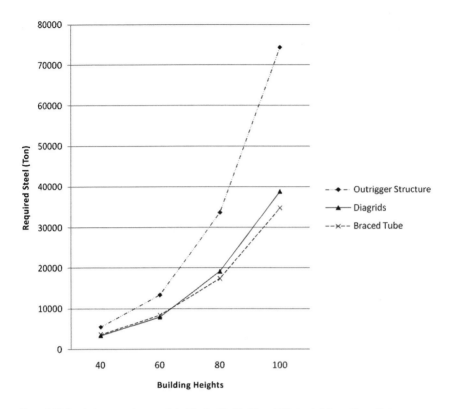

Figure 4-53. Required amount of structural steel for the 40-, 60-, 80- and 100-story buildings of braced tube, diagrid and outrigger structures.

For the 40- and 60-story buildings, the height-to-width aspect ratio of which is smaller than about 7, the study results show that the diagrid structure is the most efficient lateral load resisting system among the three systems studied. The braced tube is also very efficient for tall buildings of this aspect ratio range. For the 80- and 100-story buildings, the height-to-width aspect ratio of which is larger than about 7, the braced tube is the most efficient lateral load resisting system among the three systems studied. The diagrids are also very efficient for tall buildings of this aspect ratio range. Though outrigger structures are efficient structural systems for tall buildings in general, their lateral efficiency relying largely on the interior core structure lags behind that of braced tubes or diagrids with large perimeter diagonals.

As buildings become taller, there is a "premium for height" due to lateral loads and the demand on the structural system exponentially increases. Figure 4-53, which shows the required amount of structural steel for the 40-, 60-, 80- and 100-story buildings of braced tube, diagrid and outrigger structures, clearly illustrates this phenomenon. In the all three cases, as the building height increases, the required quantity of structural steel increases drastically. Therefore, the importance of selecting efficient structural systems becomes more significant for taller buildings. Certainly, efficiency alone is not what determines the selection of a particular structural system for a tall building. All other related design aspects should be considered holistically.

CHAPTER 5

DAMPING SYSTEMS FOR TALL BUILDINGS

THE DIRECTION OF THE EVOLUTION of tall building structural systems, based on new structural concepts with newly adopted high-strength materials and construction methods, has been towards augmented efficiency. Consequently, tall building structural systems have become much lighter than earlier ones. This direction of the structural evolution towards lightness, however, often causes serious structural motion problems – primarily due to wind-induced motion.

Tall buildings move primarily in two different directions under wind loads: along- and across-wind directions. Along-wind direction movement is intuitively clear. As the wind passes around a tall building, vortices are shed alternatively one side and then the other. This phenomenon creates alternating low pressure zones on the downwind side of the building and causes the building to vibrate perpendicular to the direction of the wind.

Tall building structures with higher lateral stiffness reduce the lateral displacement of the building. For the along-wind direction movements, this is obvious. For the vortex shedding induced across-wind direction vibrations, stiffer structures require higher wind speed for the structure to be in the lock-in condition (resonance condition), while less stiff structures can resonate even in relatively low speed wind conditions. Therefore, for both along- and across-wind direction responses, laterally stiffer tall building structures perform better in general.

In tall buildings, the lateral vibration in the across-wind direction induced by vortex shedding is generally more critical than that in the along-wind direction. For both directions, structures with more damping dissipate the vibration energy more quickly and, consequently, reduce structural

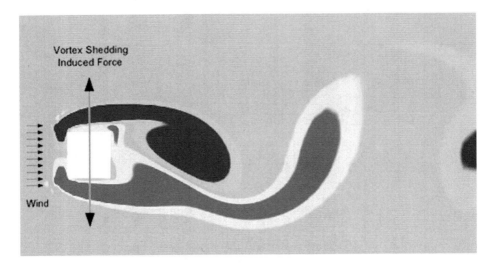

Figure 5-1. Vortex shedding induced across-wind direction force.

motions more rapidly. Since the natural direction of structural evolution towards lightness is not likely to be reversed in the future, more stiffness and damping should be achieved with a minimum amount of material.

Achievement of more stiffness in tall buildings is related to the configuration of primary structural systems, which were discussed in the previous chapter. For example, more recent structural trends such as various tubular structures including diagrids and core-supported outrigger structures achieve much higher stiffness than traditional rigid frame structures. Obtaining more damping is also related to the choice of primary structural systems and materials. However, the damping achieved by the primary structure is uncertain until the building construction is completed. A more rigorous and reliable increase in damping, to resolve tall building motion problems, could be achieved by installing auxiliary damping devices in the primary structural system. The effect of such damping can be relatively accurately estimated. Thus, when severe wind-induced vibration problems are expected, installing auxiliary damping devices can be a reliable solution.

Various damping strategies are employed to reduce the vibration of tall buildings due to wind loads. They can be divided into two categories, passive and active systems. Passive systems have fixed properties and do not require energy to perform as intended, while active systems do need an actuator or active control mechanism relying on an energy source to modify the system properties against ever-changing loads. Thus, active systems are, in general, more effective than passive systems. However, due to their economy and reliability, passive systems are more commonly used than active systems in building structures. Different types of auxiliary damping systems are summarized in Figure 5-2.

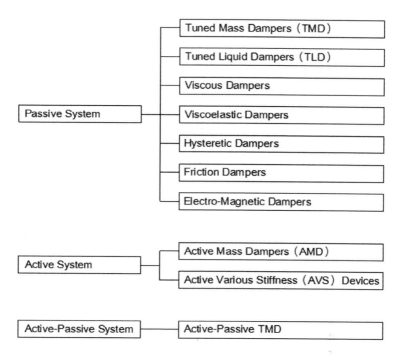

Figure 5-2. Passive and Active Damping Systems.

Seismic loads are another important lateral load to be considered for the structural design of any building. The reason why wind loads instead of seismic loads typically govern the structural design of tall buildings is related to the fundamental vibration periods of buildings, forcing periods depending on different loading types, and the impact of the resonance condition based on these two factors. The fundamental period of a building could be very roughly estimated as the number of stories of the building divided by 10. For example, fundamental periods of 10-, 50-, 100-story buildings could be estimated as 1, 5 and 10 seconds, respectively. Again, these are very rough estimations. The actual fundamental period of a building is dependent on various factors, such as slenderness ratios, structural systems and building materials.

Based on the rough estimates, the fundamental periods of low-rise buildings of 1 to 10 stories could be estimated as 0.1 to 1 second. It is not unreasonable to estimate the periods of between 0.1 and 1 second for seismic forces. Therefore, buildings lower than about 10 stories are very vulnerable to the resonance conditions with seismic forces and, consequently, serious damages or collapses. However, a 50-story building, the fundamental period of which could be estimated as about 5 seconds, is not vulnerable to the resonance condition with seismic forces of typically much shorter forcing periods.

Certainly, it is possible for the higher mode vibrations of a tall building to be in the resonance conditions with the seismic forces because the higher mode vibration periods are much shorter than the fundamental period. For example, the second and third mode periods of a 50-story building, the fundamental period of which is about 5 seconds, could be shorter than about 2 seconds and 1 second, respectively. However, the structural impact of the higher mode resonance conditions is much smaller than that of the fundamental mode resonance condition because of the characteristics of the higher mode shapes. Nonetheless, structural design of tall buildings could be controlled by seismic loads instead of wind loads in the countries with very severe seismic zones, such as Japan, Indonesia and Mexico. And some of the passive and active damping systems are sometimes employed as the primary means of seismic energy dissipation for tall buildings in these countries.

5.1. PASSIVE DAMPING SYSTEMS

The passive damping system can be further divided into two sub-categories: 1) energy-dissipating-material-based damping systems such as viscous dampers and viscoelastic dampers, and 2) auxiliary mass systems to generate counteracting inertia forces such as tuned mass dampers (TMDs) and tuned liquid dampers (TLDs).

5.1.1. Energy-Dissipating-Material-Based Damping Systems

Energy-dissipating-material-based damping systems are generally installed as integral parts of primary structural systems at vantage locations, reducing the dynamic motion of tall buildings. The damping force in a viscous damper or viscoelastic damper is dependent on the time rate of change of the deformation. Damping is accomplished through the phase shift between the force and displacement. The performance of viscoelastic and viscous dampers is dependent on the vibration frequency and temperature due to the characteristics of the materials.

Viscoelastic dampers are very effective for the vibrations of high frequencies and low magnitudes. Viscoelastic dampers, which increase both damping and stiffness, can be conveniently installed as part of diagonal braces. Typical viscoelastic dampers are composed of viscoelastic material layers bonded between steel plates, and the energy dissipation occurs with the relative movement-induced shear deformation of the viscoelastic material. The destroyed World Trade Center (WTC) Twin Towers in New York had viscoelastic dampers to resolve vibration problems. In the WTC Twin Towers, about 20,000 viscoelastic dampers were located between the perimeter columns and lower chords of the floor trusses. In the Columbia Seafirst Building in Seattle, viscoelastic dampers are incorporated with diagonal bracing members in the building core.

While viscoelastic dampers typically use the action of solids, viscous fluid dampers employ fluids to achieve passive motion control. Various different configurations of viscous fluid dampers include cylindrical pot, damping wall and piston type dampers. The cylindrical pot and wall type viscous dampers use viscous fluids in open containers. In these dampers, the piston and damping wall hung from the upper floor move through the viscous fluids based on the inter-story movements, and energy dissipations occur through the conversion of mechanical energy to heat as the piston and damping wall deform the viscous fluids. The piston type uses viscous fluids within a closed container and is more efficient than the cylindrical pot and damping wall type. In this type, the piston designed with orifices not only deforms the viscous fluid, but also force the fluid to pass through the small orifices. An example of viscous dampers, installed as an integral part of the bracing members as pistons filled with viscous fluids, can be found in the 55-story Torre Mayor in Mexico City. The viscous dampers in this building located in a severe seismic zone are used as the primary means of seismic energy dissipation.

Figure 5-3. Torre Mayor in Mexico City (left) and viscous dampers as integral part of the bracings in Torre Mayor (right). With permission of Terri Boake.

Other types of damping systems in which the damping mechanism is through direct dissipation of energy from the system include hysteretic damping and friction damping. Hysteretic damping dissipates energy through inelastic deformation of metallic substances. Various configurations of hysteretic dampers include triangular plate hysteretic dampers, buckling restrained brace frames, eccentrically braced frames, etc. In triangular plate dampers, which are typically incorporated with chevron braces, triangular plates are inserted between the chevron braces and horizontal beam element. The triangular plates are designed as the element which yields first to perform as hysteretic dampers when a large cyclic excitation occurs. In buckling restrained brace frames, concentric brace members are encased by metal jackets and the space between the braces and jackets are filled with non-bonding materials such as concrete and mortar. Therefore, the braces can be designed as yielding elements without buckling even in compressive modes. In eccentrically braced frames, the braces are placed eccentrically with typically orthogonal building frames. The short link beams created by eccentricity are designed as yielding elements to perform as hysteretic dampers.

Friction dampers adopt the concept of automotive brakes which use friction between two solid bodies to dissipate energy. X-braced friction dampers proposed by Pall and Marsh have been used in many buildings in seismic zones. The X-braced friction dampers are sometimes exposed due to their aesthetic appeal as is the case with the McConnel Library Building at Concordia University in Montreal.

5.1.2. Auxiliary Mass Dampers

Auxiliary mass dampers, such as a tunes mass damper (TMD) or tuned liquid damper (TLD), are typically installed near the top of tall buildings, where the lateral displacements are large, to generate counteracting inertia forces effectively. A TMD is composed of a counteracting-inertia-force-generating large mass accompanying relatively complicated mechanical devices that allow and support the intended performance of the mass. The frequency of the TMD mass is generally tuned to the fundamental frequency of the primary structure. Thus, when the fundamental mode of the primary structure is excited, the TMD mass oscillates out of phase with the primary structure (about one quarter of a full cycle difference between the two for optimal performance), generating counteracting inertia force.

The performance of TMDs is largely dependent on the TMD to building mass ratio. Larger auxiliary masses added to the building can produce greater counteracting inertia forces. Large TMD masses are generally made of dense materials such as steel or concrete. The 660 tonne steel and 366 tonne concrete TMD masses are employed for the 101-story Taipei 101 and 59-story Citicorp Building respectively. These large additional masses are added to the building for the purpose of damping.

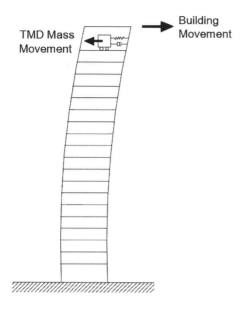

Figure 5-4. Concept of Tuned Mass Damper.

Figure 5-5. Pendulum type TMD in Taipei 101.

TMDs are generally installed in a room that is usually not accessible to the public, as in the cases of the sliding type TMDs installed in the Citicorp Building in New York and the John Hancock Building in Boston. In Taipei 101, however, the pendulum type TMD shown in Figure 5-5 is placed in the exposed five-story tall atrium-like space and performs as an ornamental element of the building as well. This ornamental installation of the TMD is unprecedented. A 6 m diameter steel ball of 660 tonnes is hung from Level 92. The steel ball is placed between Level 87 and 89. The connections between the steel

Figure 5-6. Crystal Tower in Osaka. With permission of Terri Boake.

ball and the four sets of cables hung from Level 92 are made about at Level 88. The length of the cables about four stories tall is determined to make the vibration period of the pendulum similar to the fundamental period of the building. Then, the movements of the pendulum become out of phase with the building to generate counteracting inertial forces. According to the damper designer, the damper is expected to reduce the tower's peak vibrations by more than one-third. As a building becomes taller, its natural vibration period becomes longer, and consequently the length of the cable should be longer unless compound pendulum type TMDs are used.

It is also possible to use existing masses in the building, such as ice thermal or water tanks, as TMD masses in order not to add additional masses to the building, and consequently to save cost and space. The 31-story Rokko-Island P&G Building in Kobe and 37-story Crystal Tower in Osaka use ice thermal tanks of 540 and 270 tonnes respectively. The 36-story Sea Hawk Building in Fukuoka uses 132 tonne capacity water tank as a TMD mass. Through innovative systems integrations, the intended performance can be obtained more efficiently. Compared with TMD masses made of steel or concrete, however, those primarily composed of water require much greater volume because of the much lower density of water in order to produce the identical level of damping.

Tuned liquid dampers (TLD) use waving water mass as a counter-acting inertia force generator. Two different types of TLDs used today are tuned sloshing dampers (TSD) and tuned liquid column dampers (TLCD). The former uses water typically in either rectangular or circular containers,

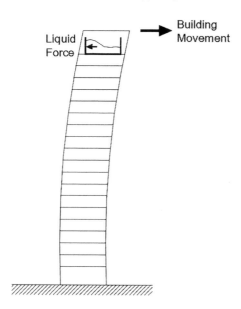

Figure 5-7. Concept of Tuned Liquid Damper.

the latter, in a U-shaped vessel. Since tall buildings have a large amount of contained water for various reasons, the TSD system can be designed using the existing water source, such as a swimming pool or water tank located near the top of the building. Therefore, an effective structural motion control mechanism can be obtained very economically, and the system is very easy to maintain. In a TSD, sloshing frequencies are tuned by adjusting the dimensions of the water container and the depth of water. The TSD system is divided into deep and shallow configurations. In the deep water configuration, the entire water mass often does not participate in providing counteracting inertia force, and the system requires baffles or screens to increase its effectiveness

Figure 5-8.
Shin-Yokohama Prince Hotel
in Yokohama.

of energy dissipation. The TSD system of shallow water configuration dissipates energy through viscous action and wave breaking.

A TSD system of shallow water configuration was installed in the 190 m tall Shin-Yokohama Prince Hotel in Yokohama. The system is composed of multiple TSD units located around the periphery of the upper floor. Each unit is composed of nine stacked layers of circular shape containers. The diameter of the containers is 2 m and the depth of water on each layer is 12.4 cm.

A TLCD can be constructed by filling liquid in a U-shaped vessel, which may be configured to work in either one or two directions. It is convenient to optimally tune the TLCD because the fundamental frequency of the damper is dependent on the length of the column of the liquid. The TLCD can also be easily incorporated with active control mechanism.

An example of TLCD installed in tall buildings can be found in the 48-story One Wall Center in Vancouver, Canada. The use of the TLCD was decided because the system can be designed to provide dual functions. A large amount of water of about 600 tonnes at the top of the building required for

Figure 5-9. One Wall Center in Vancouver and its section drawing showing TLCD at the top. With permission of Terri Boake (L), Glotman Simpson Consulting Engineers and Wall Financial (R).

damping purposes can also be used for fire suppression in the event of fire emergency. It was initially required by fire officials to install a high capacity water pump and emergency generator. However, with the installation of the TLCD, this requirement was withdrawn. Consequently, the TLCD turned out to be a very economical solution.

The 975 ft (297 m) tall 58-story Comcast Center is the tallest building in Philadelphia. In order to solve the expected lateral vibration problem of the

Figure 5-10.
Comcast Center in Philadelphia. With permission of Marshall Gerometta, CTBUH.

building, a very large TLCD, with a water mass of 1180 tonnes (300,000 gallons), was installed at the top of the building. Considering that the vibration problem occurs primarily in one direction of the larger slenderness ratio, a U-shaped uni-axial TLCD was installed.

With the height-to-width aspect ratio of 12:1, One Madison Park in New York is an extremely slender building. The 621 ft (186.2 m) tall residential tower was built on the 59 ft x 58 ft (18 m × 17.7 m) site. Reinforced concrete shear walls configured in a cruciform are employed as a primary lateral load resisting system. Because of the extremely small site and corresponding building footprint, the entire depth of the building had to be used as the structural depth against lateral loads. A perimeter tube type structural system would have substantially limited the façade design providing good views. The

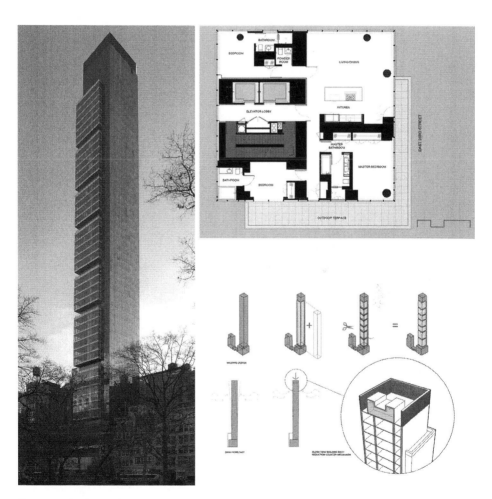

Figure 5-11. One Madison Park in New York, its typical floor plan, and its design concept with TLCD at the top. With permission of John W. Cahill (L), © CetraRuddy Architecture (RT&RB).

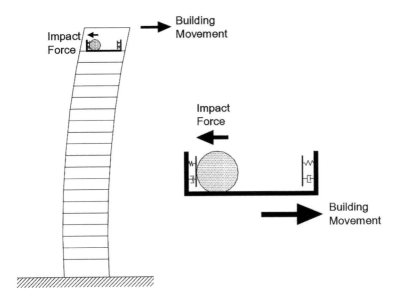

Figure 5-12. Concept of impact damper.

shear walls of the cruciform configuration allowed the façade design with good views including open corners. With the aspect ratio of 12:1, however, the tower was expected to produce vibration problems. In order to resolve this problem, TLCDs were installed at the top of the building. The system is composed of three large concrete water tanks of a U-shaped section.

Another form of inertial mass damping system is the impact dampers (ID). An impact damper is typically composed of a suspended mass within a container. The container is sized to have an optimal spacing between the mass and the container wall so that collisions occur between them as the primary structure vibrates. The frequency of the system is determined by the suspension length and the size of the mass. Though the impact damper has not been employed for any major tall buildings, it has potential as an effective motion control device.

5.2. ACTIVE DAMPING SYSTEMS

Connor defines the active structural control system as "one that has the ability to determine the present state of the structure, decide on a set of actions that will change this state to a more desirable one, and carry out these actions in a controlled manner and in a short period time" in his book *Introduction to Structural Motion Control*. Passive systems, such as TMDs or TLDs, are effective only for a narrow range of loading conditions because once the systems

Figure 5-13. Applause Tower in Osaka. With permission of Terri Boake.

are tuned for a specific target excitation they cannot adjust themselves to any untargeted variations. However, active systems, as a more advanced form of functional performance-driven technologies, can perform effectively over a much wider range of loading conditions by incorporating active control mechanisms. In active systems, control forces to adjust the system to any variations in the parameters of the system or the loading characteristics are determined by the measured response of the structure or the measured external excitation or the both. Examples are active mass dampers (AMD), active variable stiffness devices (AVSD), etc.

The AMDs resemble the TMDs in appearance except that the actuator operates on the secondary damping mass in AMDs. The vibration of a building is picked up by the sensor in AMD systems, and the optimum vibration control power calculated by a computer is generated to counteract the movement of the building. The AMDs, which have superior efficiency, require smaller masses compared with passive TMDs. Further, the AMD systems can also use existing building masses as damping masses. For example, the Sendagaya INTES Building in Tokyo and the Applause Tower (also known as Hanku Chayamachi Building) in Osaka use the 36-ton ice thermal storage tank and the 480-ton heliport, respectively, as AMD masses. However, the AMDs require higher operation and maintenance costs. In addition, since any active system requires external energy to operates, reliability is always a serious concern.

The active variable stiffness devices (AVSDs) continuously alter the building's stiffness to keep the frequency of the building away from that of external forces to avoid resonance conditions. Therefore, the AVSDs are more suitable for buildings in strong earthquake zones though the system can also be employed for tall buildings, the structural design of which is primarily governed by wind loads. The braced frames, such as the frames with inverted chevron bracings, can easily be incorporated with AVSDs. The AVSDs are attached to the bracings. The bracings, controlled by the AVSDs, are either locked to the orthogonal frame to produce a stiffer structure or unlocked to make a more flexible structure depending on the external loading conditions to avoid resonance conditions. In multistory buildings, the active controller determines which bracings to be locked or unlocked depending on the seismic ground motions in order to continuously change the lateral stiffness of the building to eliminate resonance conditions.

Although their cost-intensiveness and reliability issue is limiting the use of active systems at present, with more research, they have great potential for future applications. In fact, hybrid mass dampers (HMDs), which incorporate both TMD and AMD, can be devised to overcome the limitations of both the active and passive systems. The HMDs normally operate as passive TMDs, and their AMD mechanism is used only in the case of high excitations. However, in the event of a power failure, the system will automatically switch to the passive TMD mode. In addition, in the case of extreme excitations which exceed the capability of the actuator, the HMD system will also switch to the passive TMD mode. By this mechanism, the serious issue of reliability of AMDs can be resolved and their high operation and maintenance costs can be lowered. Further, the non-adjustability of passive TMDs can be overcome by the AMD mechanism of the HMDs in high excitation cases. The very high initial cost is a still limiting issue of the HMD systems.

CHAPTER 6
INTEGRATIVE DESIGN OF COMPLEX-SHAPED TALL BUILDINGS

THE DIRECTION OF EVOLUTION of tall building structural systems has been towards efficiently increasing stiffness against lateral loads – primarily wind loads. In order to obtain the necessary lateral stiffness, introduced first were braced frames and moment resisting frames followed by tubular structures, core-supported outrigger structures, and more recently developed diagrid structures, etc. In addition to increasing lateral stiffness, the strategy of reducing the impact of wind loads is also seriously considered by employing aerodynamic forms.

The inherent monumentality of skyscrapers resulting from their scale makes their architectural expression very significant in any urban context where they soar. While the early design of tall buildings culminated with the dominance of the International Style, which prevailed for decades and produced prismatic Miesian style towers all over the world, today's pluralism in architectural design has generated tall buildings of many different forms, including more complex forms such as twisted, tilted, tapered and free forms. This chapter presents the dynamic interactions between the various complex building forms and structural design of tall buildings.

6.1. AERODYNAMIC FORMS

In conjunction with increasing lateral stiffness against winds, a recent trend in tall building design practice is to improve aerodynamic properties of tall buildings to reduce wind forces carried by them. This can be achieved by various treatments of building masses and forms. An early example of aero-dynamic building forms can be found from Buckminster Fuller's Dymaxion project, in which the streamlined aerodynamic shield rotates about the central axis of the cylindrical form multistory building according to the direction of wind to minimize the impact of the wind force.

Figure 6-1.
Shanghai World Financial Center with a large through-building opening. With permission of Terri Boake.

Figure 6-2. Pearl River Tower in Guangzhou. With permission of Terri Boake.

Examples employed in contemporary tall buildings to improve aero-dynamic properties include chamfered or rounded corners, notches, stream-lined forms, tapered forms and openings through buildings. The Shanghai World Financial Center in Shanghai and the Kingdom Center in Riyadh employ large through-building openings at the top combined with tapered forms. The Guangzhou Pearl River Tower's funnel form façades catch natural wind not only to reduce the building motion but also to generate energy using wind turbines. Aerodynamic forms generally reduce the along-wind response as well as across-wind vibration of tall buildings caused by vortex-shedding. Due to the nature of the strategy which manipulates building masses and forms, this approach blends fittingly with architectural aesthetics.

6.2. TWISTED TALL BUILDINGS

Employing twisted forms for tall buildings is a recent architectural phenom-enon. Twisted forms employed for today's tall buildings can be understood as a reaction to rectangular box forms of modern architecture. In fact, this contemporary architectural phenomenon is not new in architecture. It is com-parable to twisted forms of Mannerism architecture towards the end of Renaissance period. For example, in Cortile della Cavallerizza at Palazzo Ducale in Mantua, Giulio Romano designed twisted columns. This twisted form can be found again in today's tall building designs such as the Shanghai Tower in Shanghai designed by Gensler, Cayan Tower in Dubai by SOM, and Chicago Spire project in Chicago by Calatrava.

In terms of static response, twisted forms do not provide any structural benefit. If solid sections are considered, the moment of inertia of a square plan is the same regardless of its twisted angle. Therefore, there is no lateral stiffness change. Building type structures are different from the solid struc-tures. They are typically composed of numerous frame elements, and their structural behavior is very much dependent on the building forms and arrangements of the frame elements. If the building type frames are considered, the lateral stiffness of the twisted forms is generally not as large as that of straight forms. This section presents studies on performance of various structural systems employed for twisted tall buildings.

The plan dimensions of the studied buildings are 36 m x 36 m with an 18 m x 18 m central core, and their typical story height is 3.9 m. The studied rates of twist are 1, 1.5, 2 and 3 degrees per floor. Stiffness-based design is performed for each structure to meet the target maximum allowable lateral displacement of a five hundredth of the building height. The ASCE document, Minimum Design Loads for Buildings and Other Structures, is used to establish the wind load and the buildings are assumed to be in Chicago.

Figure 6-3. Shanghai Tower.

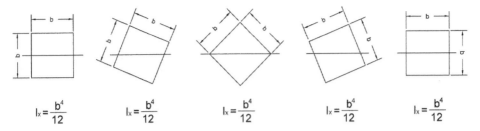

Figure 6-4. Moment of inertia of square area.

6.2.1. Twisted Braced Tubes and Diagrids

As presented in Chapter 4, both the diagrids and braced tubes are very efficient structural systems for tall buildings of conventional shapes, such as rectangular box form towers, because these systems carry lateral loads primarily by axial actions of the vertical and diagonal members on the perimeter. When these structural systems are employed for twisted tall buildings, the systems' structural efficiency is decreased as the rate of twist is increased.

Suppose the diagrids are initially configured with diagonals placed at a near-optimal uniform angle for a straight rectangular box form building.

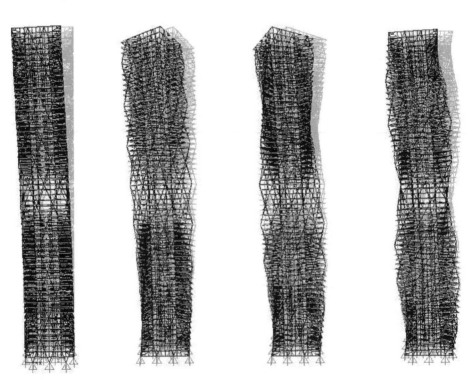

Figure 6-5. Sixty-story twisted diagrids with twisted rates of 0, 1, 2 and 3 degrees per floor.

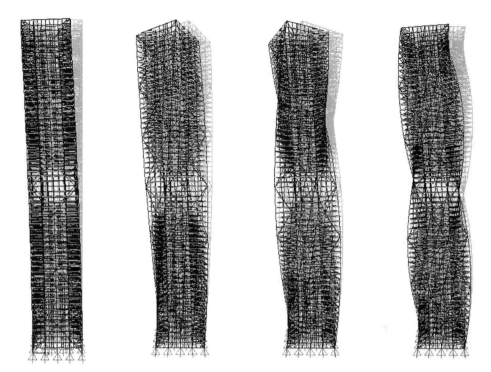

Figure 6-6. Sixty-story twisted braced tubes with twisted rates of 0, 1, 2 and 3 degrees per floor.

As the rate of twist is increased, the diagrid angle deviates more and more from its original near-optimal condition and the lateral stiffness of the system is gradually decreased. Consequently, the tower's lateral displacement is increased.

For the conventional rectangular box form towers, braced tubes are typically designed with vertical perimeter columns and diagonal bracings, which primarily carry overturning moments and shear forces, respectively, by axial actions. The vertical perimeter columns provide the maximum bending rigidity for braced tubes. As the tower begins to be twisted, the vertical columns become slanted ones. As the rate of twist is increased, slanting of the vertical columns becomes greater, which consequently reduces the system's bending stiffness gradually. Therefore, the lateral displacement of the twisted braced tube is increased as the rate of twist is increased. The angle of the perimeter diagonals is also changed by twisting the braced tube. However, the impact of diagonal angle changes caused by twisting the tower at the studied rates of 1, 2 or 3 degrees per floor is not substantial. The stiffness reduction of braced tubes, composed of verticals and diagonals, is much more sensitive to the rate of twist, compared with that of diagrids, composed of only diagonals. And this sensitivity becomes accelerated as the building height is

235

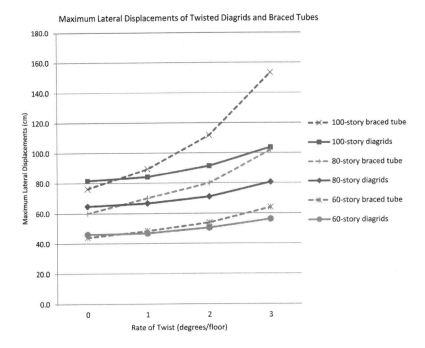

Figure 6-7. Maximum lateral displacements of twisted diagrids and braced tubes.

increased. Figure 6-7 shows maximum lateral displacements of twisted braced tube and diagrid structures of 60, 80 and 100 stories.

6.2.2. Twisted Outrigger Structures

Lateral load-carrying mechanism of outrigger structures is different from that of tube type structures, i.e., diagrids or braced tubes. Perimeter mega-columns connected to the stiff braced core structure through outrigger trusses significantly contribute to the bending rigidity of the outrigger structure. As the outrigger structures are twisted, the mega-columns on the building perimeter wrap around the building spirally. Therefore, the position of the mega-columns on the flange planes (i.e., planes perpendicular to wind) at the base changes to those on the web planes (i.e., planes parallel to wind) or even to those on the opposite flange planes at higher levels depending on the rate of twist. Lateral stiffness of the outrigger structures with these spirally slanted perimeter mega-columns is substantially reduced as the rate of twist is increased. Figure 6-9 shows maximum lateral displacements of twisted out-rigger structures of 60 and 90 stories with rates of twist of 1.5 and 3 degrees per floor.

236

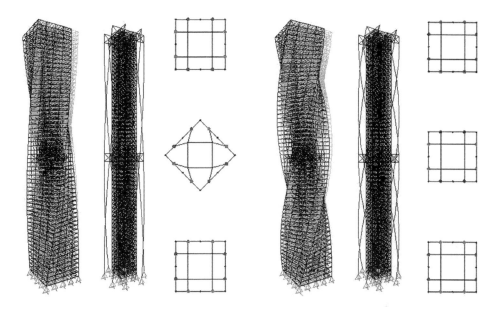

Figure 6-8. 60-story twisted outrigger structures with twisted rates of 1.5 and 3 degrees per floor.

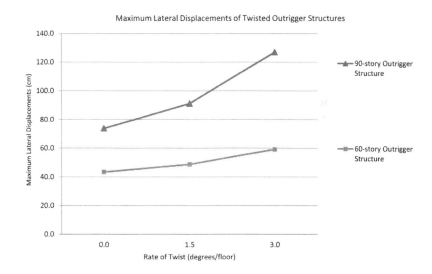

Figure 6-9. Maximum lateral displacements of twisted outrigger structures.

Outrigger structures with set-back vertical mega-columns may be a feasible alternative for twisted buildings. Compared with the straight tower case, the vertical mega-columns can be set back from the building perimeter by a certain distance so that every mega-column can be vertical rather than slanted within the rotated floor slabs. The proposed twisted form of Calatrava's Chicago Spire is designed to be supported by an outrigger system with set-back vertical mega-columns. Though the twisted outrigger buildings with set-back vertical mega-columns produce obviously smaller lateral stiffness compared with the straight outrigger buildings with perimeter mega-columns, this reduction could be smaller than that caused by using spirally slanted perimeter mega-columns. Some important design considerations related to this structural configuration include that this results in interior mega-columns between the building façades and core walls. Therefore, this configuration is more appropriate for building types which naturally require demising walls such as hotels or condominiums rather than column-free offices.

With regard to the across-wind direction dynamic responses due to vortex shedding, it should be noted that a twisted tower generally performs better than a prismatic one, as it can mitigate wind-induced vibrations by disturbing the formation of organized alternating vortexes. Considering the fact that the vortex-shedding-induced lock-in phenomenon – resonance condition

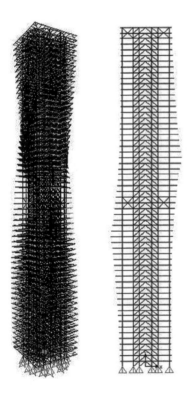

Figure 6-10.
Twisted outrigger structure with setback vertical mega-columns.

– often produces the most critical structural design condition for tall buildings, the structural contribution of the twisted building's form can be significant.

6.3. TILTED TALL BUILDINGS

Buildings have traditionally been constructed vertically, orthogonal to the ground. When a building is found to be tilted, it is typically an indication of some serious problems. The leaning Tower of Pisa is a famous example of tilted buildings due to differential settlements. Today, however, tilted buildings are intentionally designed to produce more dramatic architecture, as is the case with the Gate of Europe Towers of 1996 in Madrid designed by Philip Johnson/John Burgee, Veer Towers of 2010 in Las Vegas by Helmut Jahn, and the proposal of the Signature Towers for Dubai by Zaha Hadid. Tall buildings carry very large gravity and lateral loads. Therefore, structural impacts of tilting tall buildings are significant, and more careful studies are required for the design of tilted tall buildings. Though not uncommon these days, design and construction of tilted tall buildings is still a very recent architectural phenomenon.

In order to illustrate the concepts underlying the structural behavior of tilted tall buildings, 60-story towers of various angles of tilt are designed with

Figure 6-11. Gate of Europe Towers in Madrid. With permission of Terri Boake.

Figure 6-12. Signature Towers in Dubai. With permission of Zaha Hadid Architects.

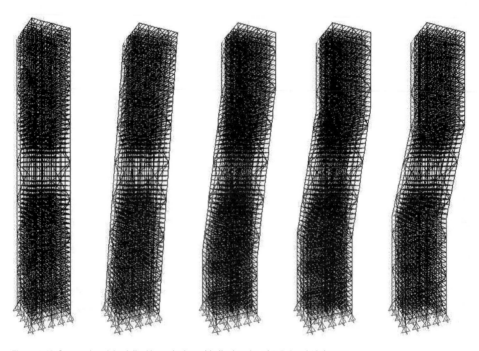

Figure 6-13. Structural models of tilted braced tubes with tilted angles of 4, 7, 9 and 13 degrees.

three different structural systems prevalently used for today's tall buildings, i.e., braced tubes, diagrids and outrigger structures. Structural steel is used for the design of all three structural systems for the straightforward comparisons, though reinforced concrete or composite structures are also commonly used in real world. Each system's structural performance depending on various angles of tilt is presented comparatively based primarily on lateral stiffness. Preliminary member sizes for the straight tower are generated first to satisfy the maximum lateral displacement requirement of a five hundredth of the building height. In order to study the structural performances of tall buildings of various tilted angles comparatively, the member sizes used for the straight structures are also used for the tilted structures for preliminary designs.

6.3.1. Tilted Braced Tubes

A 60-story tall rectangular box form straight tall building is designed with the braced tube system first, and the building is tilted at four different angles as shown in Figure 6-14. The first is the straight braced tube tower. The building's typical plan dimensions are 36 m x 36 m with an 18 m x 18 m core at the center, which produces a floor depth of 9 m between the building façades and the core perimeter walls. Typical story heights are 3.9 m. The braced tube system on the building perimeter is designed to carry the entire lateral loads, and the 18 m x 18 m building core is designed to carry only gravity loads, in order to estimate the impact of different angles of tilting on the performance of the perimeter braced tube.

Figure 6-15, with simplified section drawings of the tilted braced tubes, clearly explains the relationship between the vertical building core and the tilted perimeter braced tube for each tilted case. The second model is a tilted case with no floor offset. While the 18 m x 18 m gravity core is maintained vertical within the tilted perimeter braced tube, the building is tilted to its maximum angle of 4 degrees. Therefore, on the left side of the building as seen in Figure 6-15, the distance between the exterior façade and the core perimeter wall reduces from 18 m on the ground to 0 m at the top. On the right side, this distance increases from 0 m on the ground to 18 m at the top. Though this specific configuration produces some architectural issues regarding the space use as the distance between the exterior façade and the core perimeter wall nears 0 m, this study assumes architectural issues can be reasonably resolved in the end. The tilted form of this case is similar to that of the Gate of Europe Towers in Madrid shown in Figure 6-11 or the Veer Towers in Las Vegas.

The third, fourth and fifth models are tilted braced tube towers with floor offsets of 12, 16 and 20 stories at both the top and bottom, resulting in tilted angles of 7, 9 and 13 degrees, respectively. Tilted forms of these cases are similar to those of the Signature Towers shown in Figure 6-12. In these cases, the 18 m x 18 m gravity cores are still maintained vertical within the perimeter braced tube structures.

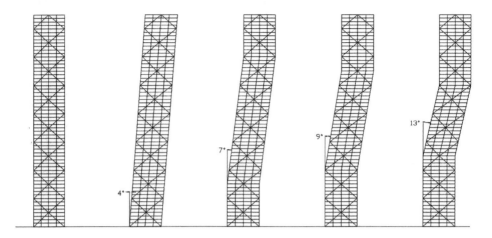

Figure 6-14. Elevations of tilted braced tubes with tilted angles of 0, 4, 7, 9 and 13 degrees.

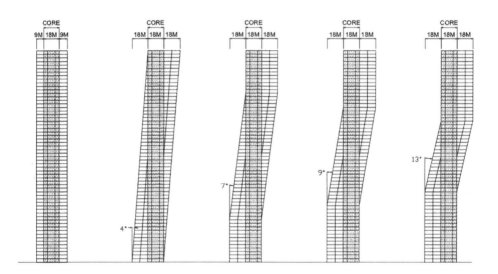

Figure 6-15. Simplified sections of tilted braced tubes with tilted angles of 0, 4, 7, 9 and 13 degrees.

Figure 6-16 summarizes the maximum lateral displacements of the tilted braced tubes in the direction parallel to the direction of tilting, when the wind load is applied also in the same direction. Lateral stiffness of the tilted braced tubes against wind loads is very similar to that of the straight braced tube regardless of the changes of the tilted angle between 0 and 13 degrees. However, initial lateral displacements of the tilted braced tubes due to gravity loads are significant. This gravity-induced lateral displacement, which is even larger than the wind-induced displacement in most cases, becomes greater

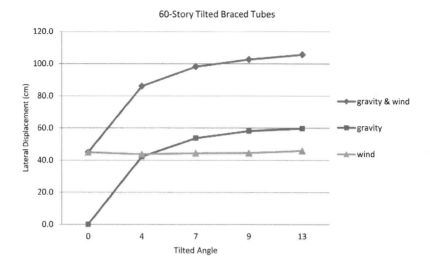

Figure 6-16. Maximum lateral displacements of 60-story tilted braced tubes.

as the angle of tilting is increased. It should be noted, however, that gravity-induced lateral displacements of tilted tall buildings can be managed substantially during construction if planned carefully.

6.3.2. Tilted Diagrids

In order to estimate the structural performance of tilted diagrids, the 60-story buildings are now designed with the diagrid structural system. Figure 6-17 shows the straight diagrid structure and its four different tilted versions. The important dimensions and tilted angles of the diagrid structures are the same as those of the braced tube structures studied in the previous section. Design conditions including applied loads are also the same as before. The major difference is that the perimeter braced tubes studied in the previous section are replaced with diagrids.

Figure 6-18 summarizes the maximum lateral displacements of the tilted diagrid towers in the direction parallel to the direction of tilting, when the wind load is also applied in the same direction. The performance of the diagrids is very similar to that of the braced tubes previously studied. Lateral stiffness of the tilted diagrids against wind loads is very similar to that of the straight diagrids regardless of the changes of the tilted angle between 0 and 13 degrees. However, initial lateral displacements of the tilted diagrids due to gravity loads are significant. This gravity-induced lateral displacement, which is even larger than the wind-induced displacement in most cases, becomes greater as the angle of tilting is increased. As has already been discussed, gravity-induced lateral displacements of tilted tall buildings can be managed substantially during construction if planned carefully.

243

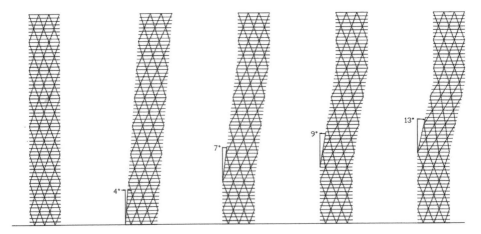

Figure 6-17. Tilted diagrids with tilted angles of 0, 4, 7, 9 and 13 degrees.

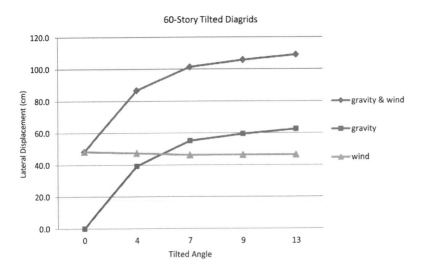

Figure 6-18. Maximum lateral displacements of 60-story tilted diagrids.

6.3.3. Tilted Outrigger Structures

In order to estimate the structural performance of tilted outriggers, the 60-story tilted buildings are now designed with the outrigger system. Figure 6-19 shows the straight outrigger structure and its four different tilted versions. The important dimensions and tilted angles of the outrigger structures are the same as those of the braced tube or diagrid structures studied in the previous sections. Other design conditions, including the applied wind loads, are also the same as before. The major differences are that the primary lateral load resisting system is changed to the outrigger system. Consequently, lateral load

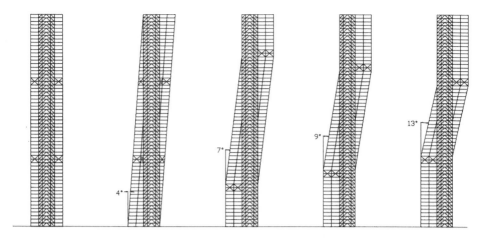

Figure 6-19. Tilted outrigger structures with tilted angles of 0, 4, 7, 9 and 13 degrees.

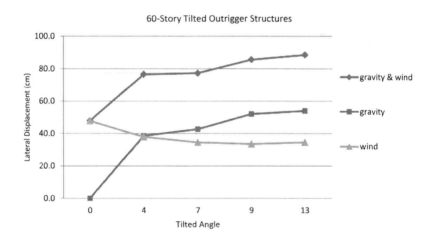

Figure 6-20. Maximum lateral displacements of 60-story tilted outrigger structures.

resisting braced frames are employed for the core structures of the outrigger systems, instead of the gravity core structures employed for the previously studied braced tubes and diagrids. For the straight outrigger tower shown in the first model of Figure 6-19, the outrigger trusses, which connect the braced frame core and perimeter mega-columns, are placed at a third and two third heights of the building. The locations of the outrigger trusses are adjusted for enhanced constructability depending on the offset locations of different cases.

Figure 6-20 summarizes the maximum lateral displacements of the tilted outrigger structures in the direction parallel to the direction of tilting,

245

when the wind load is also applied in the same direction. The performance of the tilted outrigger structures is different from that of the tilted braced tubes or diagrids. Lateral stiffness of the tilted outrigger structures against wind loads is greater than that of the straight outrigger structure. The tilted outrigger structures configured as shown in Figure 6-19 carry lateral loads more effectively because tilting the tower results in triangulation of the major structural components – the braced core, mega-columns and outrigger trusses. As the angle of tilting is increased from 0 to 13 degrees, the geometry of the triangles formed by the major structural components becomes more effective to resist the wind load, and consequently the wind-induced maximum lateral displacement of the outrigger structure is decreased. However, gravity-induced lateral displacement, which is even larger than the wind-induced displacement, still becomes greater as the angle of tilting is increased.

6.3.4. Strength Consideration for Tilted Tall Buildings

Though structural design of tall buildings is generally governed by lateral stiffness rather than strength, tilted towers are subjected to much larger localized stresses than conventional vertical towers. Figure 6-21 shows axial

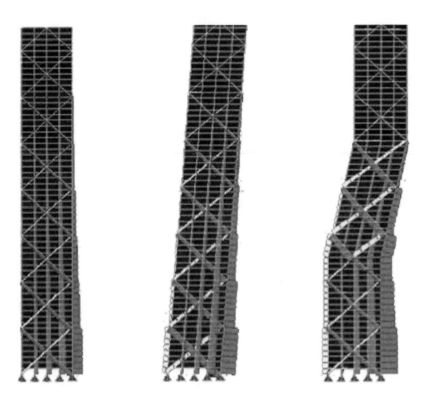

Figure 6-21. Axial member forces of the vertical and two tilted braced tube structures subjected to combined dead, live and wind loads.

member forces of the vertical and two tilted braced tube structures (the first, second and fifth models of Figure 6-14) subjected to combined dead, live and wind loads. Much larger compressive and tensile member forces are developed in the tilted braced tubes than in the straight braced tube.

Tensile forces developed in tall buildings due to wind loads are often cancelled by compressive forces caused by gravity loads. In the tilted braced tubes studied here, however, substantial tensile forces are developed in perimeter columns and bracings due to the eccentricity. More careful studies are required for the design and construction of the connections of these members and foundations.

6.4. TAPERED TALL BUILDINGS

Compared with prismatic forms, tapered forms provide many advantageous aspects for structural systems of tall buildings. The magnitudes of lateral shear forces and overturning moments become larger towards the base of the building. Consequently, greater lateral stiffness is required at lower levels. Tapered forms provide greater lateral stiffness towards the base because tapered forms naturally produce greater structural depth towards the base.

Tapered forms also help reduce wind loads applied to tall buildings. Wind pressure is greater at higher levels and lesser at lower levels due to the friction with the ground surface. When a building is tapered, the exterior surface area where the wind load is applied is reduced at higher levels, and increased at lower levels. Therefore, the lateral shear forces and overturning moments are decreased as the angle of taper is increased.

For tall buildings, vortex-shedding induced lock-in conditions often create the most critical structural design issue. Tapered forms help tall buildings prevent shedding organized alternating vortices, which cause the lock-in condition, because of the continuously varying building width along the building height. Therefore, tapered tall buildings are less susceptible to severe across-wind direction vibrations than prismatic tall buildings.

Furthermore, tapered forms are often more desirable architecturally for mixed-use tall buildings. For residential functions in tall buildings, for example, it is important to make living spaces not too far away from natural light to enhance the occupant comforts. For commercial office functions, however, daylighting is less important and deeper rentable space is often desired. Therefore, tapered tall buildings, with commercial office functions on the lower levels and residential functions on the higher levels, perform very well architecturally. A famous example of this type of spatial organization in a tapered tall structure can be found in the 100-story tall John Hancock Center of 1969 in Chicago designed by Skidmore, Owing and Merrill.

Tapered tall buildings of 60, 80 and 100 stories are designed with braced tubes, diagrids and outrigger structures and their structural perform-

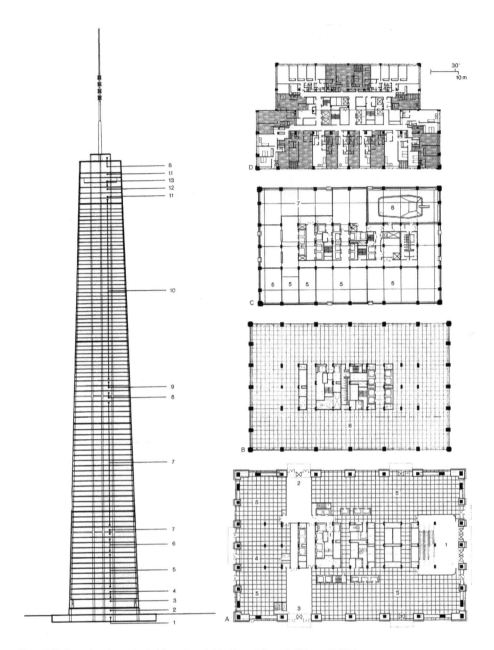

Figure 6-22. Tapered section and typical floor plans of John Hancock Center in Chicago. © SOM.

ances are studied comparatively in this section. The height-to-width aspect ratios of the studied buildings range from about 6 to 10. The angles of taper studied are 1, 2 and 3 degrees.

Figure 6-23 shows 60-story buildings of the three different angles of taper designed with braced tubes. A non-tapered straight braced tube structure is designed first. The straight building's typical plan dimensions are 36 m x 36 m with an 18 m x 18 m core at the center and typical story heights of 3.9 m. The perimeter braced tube system is designed to carry the entire lateral load, and the 18 m x 18 m building core is designed to carry only gravity loads, in order to clearly estimate the impact of taper on the performance of the perimeter braced tube system. The building is assumed to be in Chicago and the ASCE document, Minimum Design Loads for Buildings and Other Structures, is used to establish the wind load. This loading condition is also used for the diagrids and outrigger structures. The braced tube member sizes for the straight tower are determined based on the stiffness-based design methodology to meet the maximum lateral displacement requirement of a five hundredth of the building height.

The braced tube structure is then tapered with three different angles of 1, 2 and 3 degrees as shown in the figure. Since the building width at mid-height is maintained to be the same as the original straight tower, each tapered building's gross floor area is the same regardless of the different angles of taper. Member sizes determined for the straight braced tube are also used for the tapered ones for preliminary design purposes to investigate the impact of tapering the structure comparatively.

The maximum lateral displacements of the 60-story tapered braced tubes shown in Figure 6-23 as well as similarly configured and designed 80- and 100-story tapered braced tubes are summarized in Figure 6-24. As has been predicted, lateral displacements of tapered braced tubes are reduced as the angle of taper is increased. Further, the rate of displacement reduction due to taper is accelerated as the building becomes taller.

As another perimeter tube type structure with diagonals, the result of the same study with diagrid structures is very similar. As the angle of taper is increased, the lateral stiffness of the perimeter tube type structures is substantially increased, and the rate of displacement reduction due to taper is accelerated as the building becomes taller. The lateral performance charac-teristics of tapered perimeter tube type structures are very similar.

In outrigger structures, as the building is tapered while maintaining its width at mid-height, the stiffness of the lower level outrigger trusses, which connect the mega-columns and braced core, is reduced because their length is increased. This makes the lateral performance of tapered outrigger structures slightly different from that of the perimeter tube type structures. However, the lateral stiffness of outrigger structures is still increased substantially, as the angle of taper is increased, and the rate of displacement reduction due to taper is accelerated as the building becomes taller.

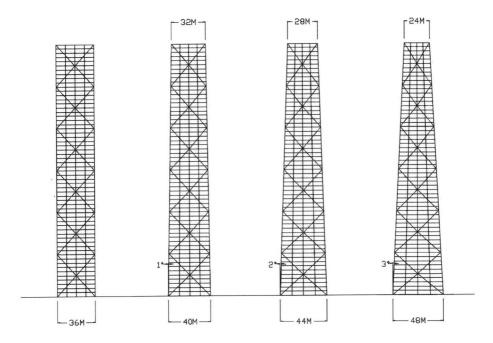

Figure 6-23. Structural models of 60-story tapered braced tubes.

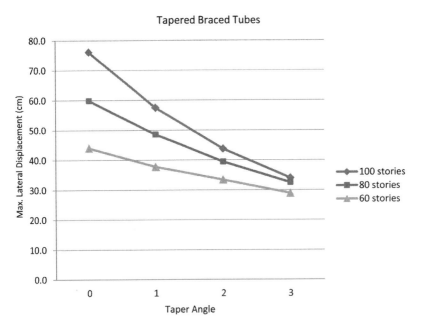

Figure 6-24. Maximum lateral displacements of 60-, 80- and 100-story tapered braced tubes.

6.5. FREEFORM TALL BUILDINGS

Another distinguished architectural design approach for today's tall buildings is sculptural freeform design. Early examples of this approach include Peter Eisenman's Max Reinhardt Haus and Frank Gehry's New York Times Building, both designed in the late 20th century but not built. Today, however, many freeform tall buildings are designed and actually built worldwide.

It was quite a difficult task to perform the structural designs and analyses of irregular freeform tall buildings in the past. These days, it can be relatively easily done with the development of structural design/analysis computer software. Even though the supporting structural systems behind the free forms vary depending on the project-specific situations, diagrids are

Figure 6-25. Max Reinhardt Haus project by Peter Eisenman (unbuilt). With permission of Eisenman Architects.

often employed as can be observed from RMJM's Capital Gate Tower.

As a building's form becomes more irregular, finding an appropriate structural system for better performance and constructability is essential to successfully carry out the project. The diagrid structural system has great potential to be developed as one of the most appropriate structural solutions for irregular freeform towers. Triangular structural geometric units naturally defined by diagrid structural systems can specify any irregular freeform tower more accurately without distortion.

Diagrid systems are employed for 60-story freeform tall buildings to investigate their structural performance in this section. Freeform geometries are generated using sine curves of various amplitudes and frequencies. For the purpose of comparison, preliminary member sizes for the 60-story rectangular box form diagrid tall building are generated first to satisfy the maximum lateral displacement requirement of a five hundredth of the building height.

Once the conventional rectangular box form diagrid structure is designed, three different freeform diagrid cases shown in Figure 6-26 are comparatively studied. To estimate the lateral stiffness of diagrid structures employed for freeform structures, the member sizes used for the straight rectangular box form tower are also used for the freeform towers. Thus, each structure is designed with very similar amount of structural materials. Compared with the rectangular box form diagrid structure, which has 36 m x 36 m square plan on each floor, the first, second and third freeform cases' floor plans fluctuate within the ± 1.5 m, 3 m and 4.5 m boundaries of the original square respectively. Despite these geometry changes, total floor area of each case is maintained to be the same.

As can be seen in the figure, which shows the deformed shape of each diagrid structure in a scale factor of 20, the lateral displacement of the structure becomes larger as the freeform shape deviates more from its original rectangular box form. This is related to the change of the diagrid angle caused by free-forming the tower. The straight tower designed first for the comparison is configured with the optimal diagrid angle of about 70 degrees. As the degree of fluctuation of freeform is increased, the diagrid angle deviates more from its original optimal condition, which results in substantial reduction of the lateral stiffness of the tower. Therefore, freeform shapes should be determined with careful consideration of their not only architectural but also structural performance.

With regard to the across-wind direction dynamic responses due to vortex shedding, it should be noted that irregular freeform towers, like previously studied twisted towers, generally perform better than a prismatic one, as it can mitigate wind-induced vibrations by disturbing the formation of organized alternating vortexes. Considering the fact that the vortex-shedding-induced lock-in phenomenon often produces the most critical structural design condition for tall buildings, the structural contribution of irregular free form can be significant.

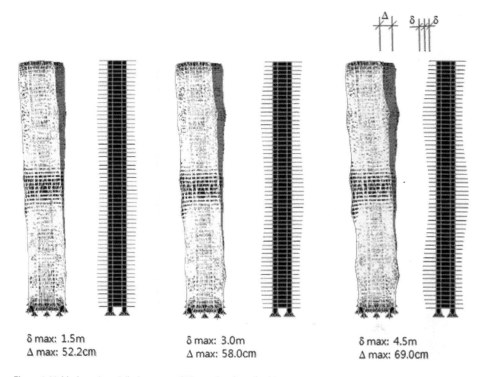

δ max: 1.5m
Δ max: 52.2cm

δ max: 3.0m
Δ max: 58.0cm

δ max: 4.5m
Δ max: 69.0cm

Figure 6-26. Maximum lateral displacements of 60-story free-form diagrids.

6.6. CONJOINED TALL BUILDINGS

In addition to the various complex-shaped tall buildings presented in the previous sections, another recently emerging tall building design strategy is conjoining two or more tall buildings functionally and structurally. Quite a few competition entries for rebuilding the World Trade Center in New York employed this approach, though the winning entry did not. In fact, the concept of conjoined towers dates back to the King's View of New York published by Moses King in the early 20th century. The King's Dream of New York in this publication shows New York skyscrapers connected by sky bridges. Today, interconnected tall buildings are no longer a dream.

The Petronas Towers in Kuala Lumpur are twin commercial office towers connected by a double story sky bridge on the 41st and 42nd floors. In the Pinnacle at Duxton in Singapore, seven residential towers are connected on the 26th and 50th floors by sky bridges. The same strategy is used in the Sky Terrace at Dawson in Singapore, which connects four residential towers on two levels. While connecting structures in these examples are simple bridges between the buildings, some other connecting structures are designed to hold substantial programed spaces. In the Marina Bay Sands in

Figure 6-27. Marina Bay Sands. With permission of JH Song.

Singapore, three hotel towers are connected at the top by the sky park that brings together many hotel amenity facilities, such as swimming pools, restaurants and sky gardens. A very similar design approach is used in the Gate Towers in Shams Abu Dhabi on Reem Island and Raffles City Chongqing in Chongqing.

A new conjoined towers typology has also been introduced in some projects. For example, the CCTV Tower in Beijing can be better conceived as a closed loop type tall building designed to produce unique and enhanced functional performance, instead of two towers connected by bridging programed spaces. The concept of closed loop type tall buildings can be found also in the proposed mixed-use Infinity by Crown Group in Sydney. By inter-connecting towers with various new design concepts, tall buildings are no longer isolated individual towers. They are growing into organically intercon-nected more dynamic megastructures functioning like vertical cities.

The concept of conjoined towers also has great structural potential to produce very tall buildings. One of the tallest proposed conjoined towers was the Nakheel Tower for Dubai. The proposed height of the tower was over 1,000 m, though the project was cancelled. The Nakheel Tower can be conceived as four megatall buildings structurally belted together at every 25

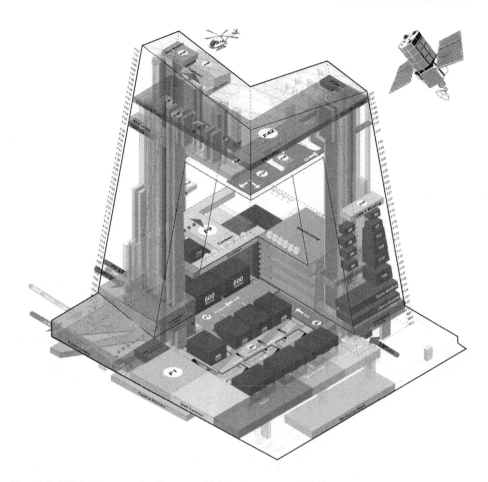

Figure 6-28. CCTV Headquarters, a closed loop type tall building. Image courtesy of OMA.

stories. This configuration can provide excellent structural performance for supertall and megatall buildings. The large openings between the four towers significantly reduce wind loads, and the large footprint of the conjoined towers produces a great structural depth against lateral loads.

Though it is very difficult to foresee what is coming next to the field of supertall and megatall design, the structurally conjoined towers concept employed for the Nakheel Tower project is expected to serve as one of the powerful prototypes for upcoming supertall towers over 1,000 m. For very tall buildings, it is crucial to maximize their structural depths to efficiently provide lateral stiffness against wind loads. At the same time, it is very important to keep reasonable lease depths – about 8–15 m for the comfort of occupants. The structural concepts developed for the Nakheel Tower in conjunction with building forms can satisfy these fundamental structural and architectural requirements successfully.

Figure 6-29. Nakheel Tower project (cancelled). Project: Nakheel Harbor and Tower, Client: Nakheel Properties, Architect: Woods Bagot, Structural Engineer: WSP Cantor Seinuk.

For extremely tall buildings, structural systems cannot be configured independently without considering building forms. If a structural depth of about 100 m is used for a 1,000 m tall conventional rectangular box form building with a central core to maintain its height to width aspect ratio of about 10, the lease depth between the façade and core wall will be very deep. This depth will be about 25 m when square floor plan and square central core are considered with the core to gross floor area ratio of about 25 percent. The height to width aspect ratio is one of the most important factors for structural design of very tall buildings. With the same slenderness and core area ratio, if the height of the rectangular box form building is increased to 1,200 m, the lease depth should also be increased to about 30 m. Extremely large lease depths for megatall buildings of unprecedented heights will produce many serious issues, such as occupant discomfort, interior columns within the lease depth to avoid floor beams of very long spans, and too large gross floor area, to list but a few.

In order to maintain the height to width aspect of about 10, the plan dimensions of the about 1,000 m tall Nakheel Tower on the ground are very large, about 100 m. However, the structurally conjoined tower concept allows this building to maintain reasonable lease depths. Further, unlike the conventional rectangular box form tall building with a central core, towers of this configuration can be designed even taller to a certain height without increasing lease depths, because increasing the overall plan dimensions to keep the slenderness and maintaining the desired lease depth can be done independently to a large degree. Figure 6-30 comparatively shows simplified structural plans of the structurally conjoined towers and conventional rectangular box form concepts employed for 1,000 m and 1,200 m tall buildings, respectively.

Despite its characteristics being appropriate for extremely tall buildings, this architecture-integrated structural concept may not be easily

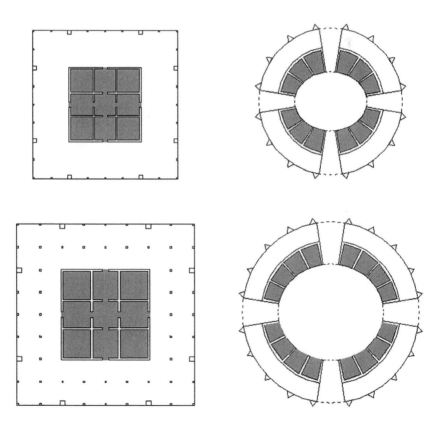

Figure 6-30. Simplified structural plans of the structurally conjoined towers and conventional rectangular box form concepts employed for 1,000 m and 1,200 m tall buildings.

employed for existing dense urban land because very large building sites are required. However, where appropriate, these conjoined towers may be the solutions for the problem of dense urban environments, by way of creating three-dimensional vertical cities in the sky.

Structural potential of conjoined towers can be further increased, and an extremely tall building complex can be designed in a more efficient way, by extending Fazlur Khan's superframe concept strategically and in an integrative way with other design aspects. Figure 6-31 shows a mile-high superframed conjoined tower design project by Chris Hyun at Yale School of Architecture, under the guidance of Kyoung Sun Moon. The project was proposed for the empty site in Chicago, partly including the area once used for the never-completed Chicago Spire project by Santiago Calatrava. In this design project, four exceedingly tall buildings are interconnected with the structural concept of the superframe, to create the mile-high conjoined towers.

Four braced-tube towers are placed in the corners of the enormous superframe, allowing it to reach the height of one mile (about 1.6 km). The braced-tube towers are connected by horizontal bands of braced tube

Figure 6-31. Mile-high superframed conjoined towers.

structures of multiple story height, which become the connections between the towers, potentially housing sky lobbies and other public spaces of truly city-like conjoined mega-towers. At the same time, these are what create the superframed conjoined towers which use the entire width of the tower complex as the structural depth, instead of the width of the individual towers. Therefore, the stiffness of the horizontal connection braced tubes is a very important structural design consideration of these towers. By structurally inter-connecting multiple towers, greater structural depths, as a group against lateral loads for an enormous height and desired lease depths for individual towers for better functional performance, can be achieved simultaneously.

INDEX

Printed and bound by PG in the USA